Planet Earth

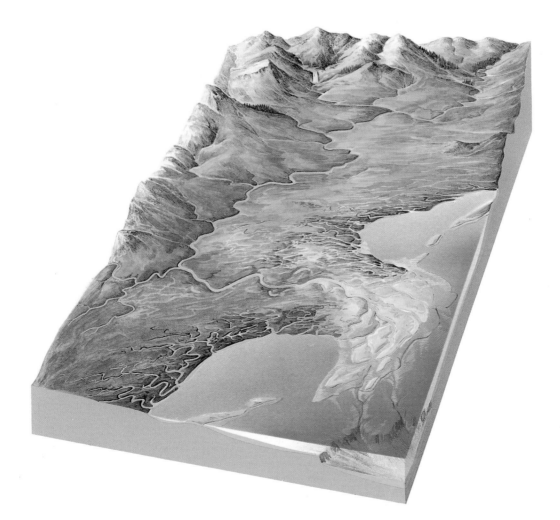

Oxford University Press

Oxford University Press, Walton Street, Oxford *OX2 6DP*

Oxford   New York
Athens   Auckland   Bangkok   Bombay
Calcutta   Cape Town   Dar es Salaam   Delhi
Florence   Hong Kong   Istanbul   Karachi
Kuala Lumpur   Madras   Madrid   Melbourne
Mexico City   Nairobi   Paris   Singapore
Taipei   Tokyo   Toronto

and associated companies in
Berlin   Ibadan

Oxford is a trade mark of Oxford University Press

© Oxford University Press 1993
First published 1993
Reprinted 1995

British Library Cataloguing in Publication Data
Data available

ISBN 0-19-910144-2

Regions of the World text by Jill Bailey
Edited by Jill Bailey and Catherine Thompson
Designed by Richard Morris, Stonesfield Design

Printed in Hong Kong

Foreword

Open *Planet Earth* and begin to explore the World. Discover how mountains and lakes are formed, and what alters the shape of the land. Find out about countries, their location, geography, population, and much more.

High-quality text and artwork from the *Oxford Children's Encyclopedia* have been adapted, and expanded with much new material, to produce this easy-to-use reference guide to *Planet Earth*.

How to use *Planet Earth*

Like all reference books you can use *Planet Earth* in two different ways. Make time to sit and browse through it for pleasure, and you will soon find yourself engrossed in a subject you would never have thought to look up.

Planet Earth is organized in five sections. The first half of the book looks at aspects of physical geography, weather, and the environment. The second half explores the World's regional geography, with a map section and numerous country fact files.

When you wish to find out about a particular subject, and have no time for browsing, using the index is the quickest way.

V

valleys 22-24, *22, 23, 24* ——— text information ——— pictures
veld 124-126, *139*
Venezuela 135-138, *140*

Say you wanted to find out about valleys, you must turn to the index, which is always at the back of a book and organized alphabetically. Under the entry **valleys** you will find the same page number twice. The first numbers 22–24 tell you that you will be able to read about valleys by turning to page 22. The second numbers *22, 23, 24* in *italics*, tell you that you will also find pictures of valleys on those pages.

Contents

Foreword

The shape of the land

Environments and weather

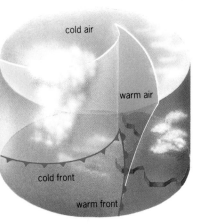

cold air

warm air

cold front

warm front

Regions of the World

Introduction *84*

World maps

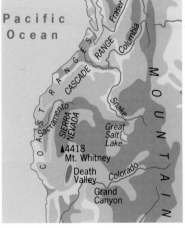

Countries at a glance

Earth

Area of Earth's surface
510 million sq km

Area of land
149 million sq km (29%)

Area of sea
361 million sq km (71%)

Volume of the Earth
1,083,230 million cubic km

Weight of the Earth
5,976 million million million tonnes

Age of the Earth
4,600 million years

Origin of life
3,500 million years

Pressure at the core
3·5 million atmospheres

Temperature at the core
5,000°C

The Earth is different from the other planets in the Solar System because it has water and an atmosphere containing oxygen, so life can exist here. The Earth goes around the Sun in a great orbit once every year. This causes the seasons. The Earth also spins round on its own axis once every 24 hours, and this causes day (when one side faces the Sun) and night (when the same side faces away). The Earth is tiny when compared with many other planets, or with the Sun. The planets Mercury, Venus, Mars and Pluto are smaller; Jupiter and Saturn are hundreds of times bigger. The Sun is over a million times bigger than the Earth.

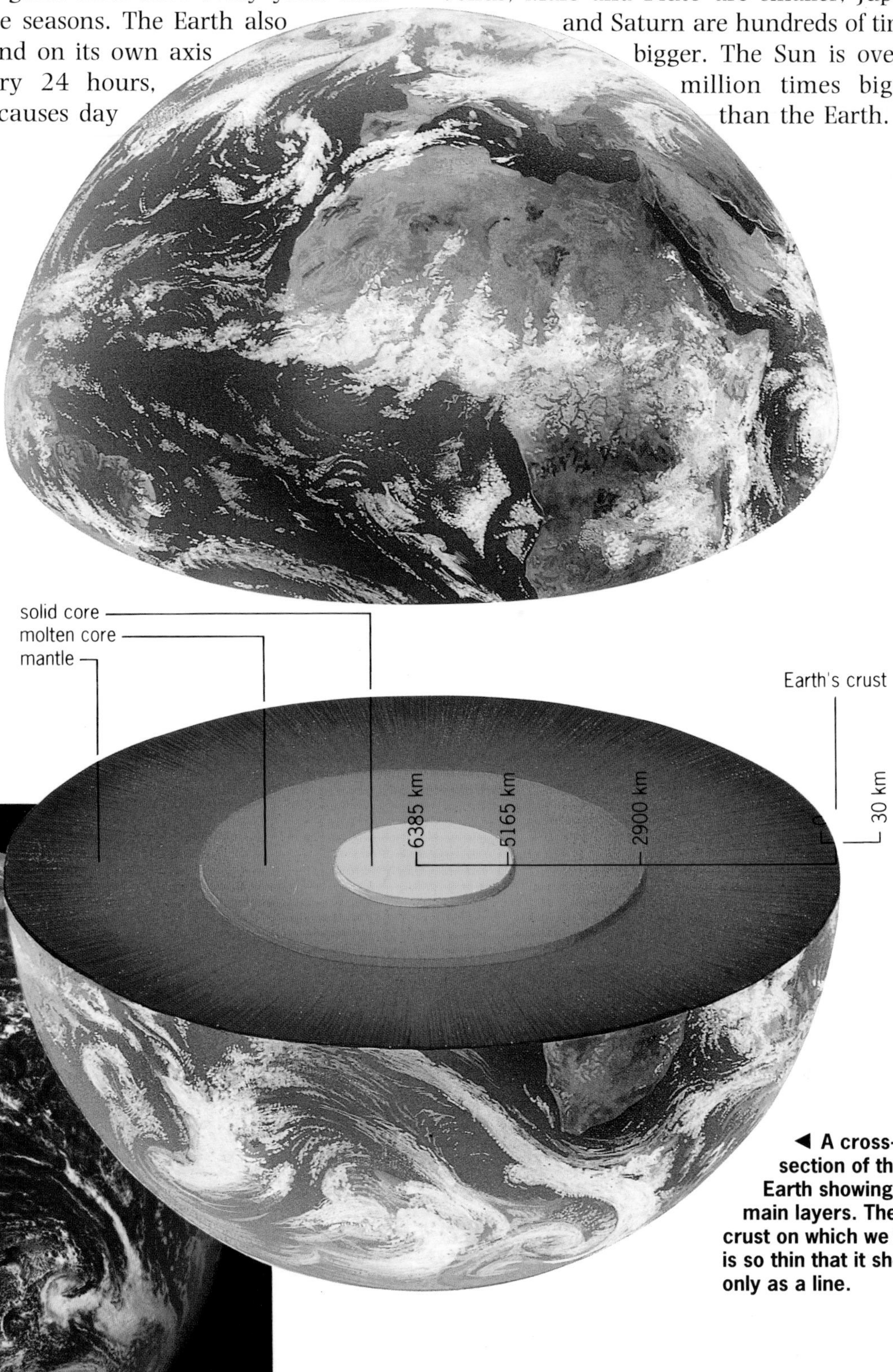

solid core
molten core
mantle

Earth's crust

6385 km
5165 km
2900 km
30 km

◄ A cross-section of the Earth showing the main layers. The crust on which we live is so thin that it shows only as a line.

CONTINENTS ON THE MOVE

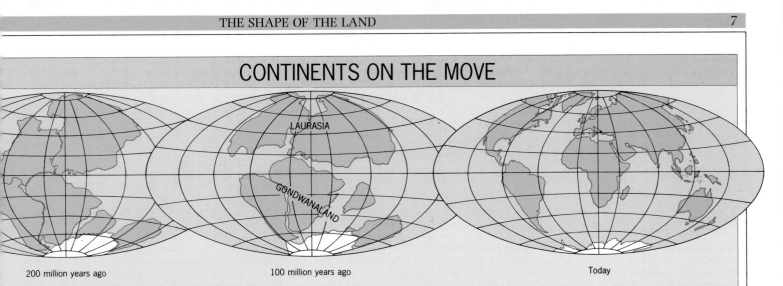

200 million years ago   100 million years ago   Today

In the distant past the Earth looked very different. The continents and the oceans have changed tremendously. For example, 200 million years ago there was no Atlantic Ocean. North America, Europe and part of Asia were joined as one continent which we call Laurasia. South America, Africa, India, Australia and Antarctica formed a large southern continent called Gondwanaland. We can work out when the Atlantic began to open up, and how wide it was at different times.

This continental drift seems evident when we see how well the coastlines match up. For example, Africa and South America fit neatly together. Also, the sands and muds on the bottom of the ocean can be dated accurately, and they become younger and younger as you get nearer to the centre of the ocean.

What is happening is that there is a deep crack right up the middle of the Atlantic which is like a line of small volcanoes. Molten rock is forced up the crack and the two halves of the Atlantic are pushed apart. A new ocean floor is created in the middle with Africa gradually shifting eastwards, South America westwards. This movement is only 1 to 10 cm (⅜ in to 4 in) per year. The surface of the Earth is divided up into several large plates, and these are all moving in different directions. North America will eventually reach Russia to the west. California (on the Pacific plate) may drift away from the rest of north America.

▲ The present-day continents began to take shape about 200 million years ago. Since then the various land masses have drifted to their present location.

Inside the Earth

Scientists do not know exactly what the Earth is like inside. We live on the outer part which is made from hard rocks and covered with water in places. This is the crust. The inside of the Earth is very hot, and below about 70 km (40 miles) the rocks are all in the form of molten liquid. We know this because miners have found that the rocks become warmer and warmer down deep mines, and molten rock often comes to the surface through volcanoes. The main inner layer, the mantle, is made of molten rock, and the inner core of the Earth is made from solid and liquid metal.

Origin of the Earth

The Universe is said to have come into existence as much as 20,000 million years ago. We shall never really know how this happened. The most popular explanation is the 'big bang' theory. According to this, there was an enormous explosion which sent gases and particles hurtling out in all directions. Gradually, galaxies and solar systems began to form from this swirling mass, and the Sun might have eventually formed about 5,000 million years ago.

The Earth is thought to have formed 4,600 million years ago as a ball of molten rock. It was probably as hot as 4,000°C. It took many millions of years for the Earth to cool down enough for a crust to form, and the crust must have been very thin at first. Molten rock turns into solid rock at temperatures of 800–1,500°C, so the early Earth's crust would have been too hot to stand on.

The early Earth probably had no atmosphere, just like most of the other planets now. However, gases were ejected from volcanoes all over the surface, and a primitive atmosphere developed. This atmosphere had no oxygen. The first forms of life are believed to date from 3,500 million years ago, and they lived without oxygen. Oxygen came much later, since it is produced mainly by plants.

Mountains

Young and old mountains

As soon as mountains start to rise above the surrounding land they are attacked by water, wind, rain and ice. The softer rocks are soon eroded to form valleys, and jagged peaks form as the valleys are cut back into the hills. Young and rapidly rising mountains have high steep-sided peaks and deep valleys and gorges. In time, the peaks of the mountains and the sides of the valleys are worn away, the hills become lower and more rounded and the valleys become wider, with slower-flowing, less powerful rivers. In old age, a mountain range may become little more than a gently undulating plain.

Mountain climate

As you climb up a mountain, the temperature falls by 1°C for every 150 m (500 ft). There is snow on the top of high

▶ Jagged, snow-capped peaks in the Cordillera range of the Andes in Argentina. The formation of the Andes began about 80 million years ago. As plates of the Earth's crust move together the mountains continue to rise.

What is the difference between mountains and hills?
Mountains have very steep slopes and peaks. The summits are usually very high. Hills are smaller and lower. But it is really a matter of judgement and comparison. What are called mountains in one region might be the same height as features that are called hills somewhere else.

HOW MOUNTAINS ARE MADE

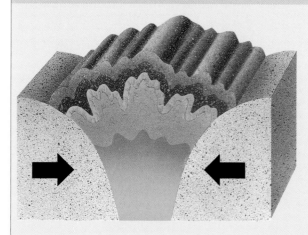

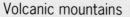

Fold mountains

The continents sit on great plates of the Earth's crust which slowly move over the surface of the planet. Where two plates move towards each other, the sediments on the floor of the ocean between them are squeezed up to form mountain ranges, such as the Alps, Himalayas, Andes and Rockies. This is why rocks containing fossils of marine animals can be found at the top of mountains.

Volcanic mountains

Many mountains are formed when molten rock from deep inside the Earth rises to the surface. It may pour out of the ground as lava, forming a volcano. As the lava cools, it forms hard, solid rock. Mount Fuji in Japan and Mount Vesuvius in Italy are volcanoes. The largest volcanic mountain in the world is Mauna Kea in Hawaii, which rises 10,000 m (over 30,000 ft) from the bed of the Pacific Ocean.

mountains even at the Equator. Also there is less oxygen in the atmosphere the higher up you go. Winds are often very strong, and the weather can change very quickly. Mountains have a dramatic effect on the climate of surrounding areas. As the clouds rise over the mountains, they shed their rain, so the side of a mountain range where the wind blows is often very wet, but the sheltered side (the rain-shadow) gets very little rain. An example of this is in California where the western side of the Sierra Nevada receives rain brought in from the Pacific, but on the eastern side is the desert called Death Valley.

Wildlife and vegetation

Mountain plants and animals have to cope with extremes of temperature: hot days and cold nights, and very high winds. Mountain soils are often thin, as soil is washed down the slopes. Above a certain altitude, called the tree-line, trees cannot grow as conditions are too harsh, and the vegetation is stunted and slow-growing. Many animals, such as mountain hares and ptarmigan, have extra thick coats in winter.

Highest mountain
Mount Everest, 8,863 m (29,079 ft).
Mauna Kea in Hawaii, although only 4,205 m (13,796 ft) above sea-level, has about another 6,000 m (20,000 ft) below water: a grand total of about 10,000 m (over 30,000 ft).

Biggest land mountain range
Himalaya-Karakoram range contains 96 of the world's 109 peaks over 7,315 m (24,000 ft) high.

Longest underwater mountain range
Indian/East Pacific Oceans Cordillera is 19,000 km (12,000 miles) long.

Block mountains

Sometimes huge blocks of rock can split and slide along lines of weakness called faults. Great masses of rock may be tilted or lifted above the neighbouring rocks to form mountains. Examples are the Sierra Nevada mountains in the western United States, the Black Forest and Harz mountains of Europe, and Mount Ruwenzori in East Africa, which reaches a height of 5,167 m (16,952 ft).

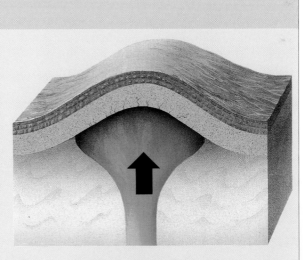

Dome mountains

Sometimes the molten lava does not reach the surface, because the rocks above are too strong to give way. Instead, it forces the rocks to bulge upwards, to form a dome-shaped mountain. Sometimes the pressure comes not from lava, but from moving water underground, thick with dissolved salts. The Black Hills of South Dakota, in the USA, are the eroded remains of a dome mountain.

Himalayas

The Himalayas cover an area of about 594,400 sq km (229,500 sq miles) and include parts of Pakistan, India, Nepal, Bhutan and Tibet.

The Himalayas are 2,500 km (1,500 miles) long and 200–400 km (125–250 miles) wide.

Highest mountains
Mount Everest
8,863 m (29,079 ft)
K2
8,611 m (28,250 ft)

Measurements of the height of Everest vary. The measurement in 1852 by the Government of India Survey Department gave a height of 8,840 m. A Chinese measurement in 1987 gave 8,848. The Research Council in Rome in 1987 announced a height of 8,863 m. Satellites were used in the last measurement.

▼ **Beyond Nuptse, Mount Everest shines in the light of the setting Sun.**

The Himalayas are a series of mountain ranges which curve in a great arc for 2,500 km (1,500 miles) from Pakistan in the west to Tibet in the east. They form the largest mountain system in the world. Thirty mountains reach heights of over 7,300 m (24,000 ft) above sea-level. There are many jagged snow-capped peaks and large valley glaciers. In places, the rivers have cut gorges up to 4,900 m (16,000 ft) deep. Nineteen large rivers drain the Himalayas, including the rivers Indus and Brahmaputra. They carry silt and mud from the eroding mountains to the great flood-plains in India and Bangladesh, forming the rich soil on which crops like rice and cotton are grown.

The Himalayas began to form about 38 million years ago. The sediments of the ancient Tethys Sea became crushed and folded as the continental plate bearing India was forced up against the Eurasian plate. The mountains are still rising today. In the heart of the Himalayas are very ancient rocks, up to 4,600 million years old.

The climate is very different on the north and south sides of the mountain system. To the south, India and Pakistan are protected from cold air from the north, and have a temperate climate. As the rain-bearing monsoon winds blow north from the Indian sub-continent, they are forced to rise over the Himalayas, dropping their rain and

snow. Parts of the southern Himalayas have over 3,000 mm (120 in) of rainfall a year. North of the Himalayas, the winds have lost their moisture. In Tibet desert conditions exist.

Rice, cereals, sugar cane and other crops are grown in the river valleys and on hill terraces. Orchards of fruit trees, vineyards and tea plantations flourish on the lower slopes. Sheep, goats and yaks, a kind of mountain cattle with long shaggy hair well-suited to the harsh mountain climate, are reared, too. Yaks are also used to carry people and heavy loads.

About one-third of the Himalayas is covered in forest, used to make paper, matches and other products. The trees also provide firewood for local people. But too many trees have been felled on the steeper slopes, and the thin exposed soil cannot absorb the heavy rainfall. The soil washes away down the slopes, and disastrous floods occur further down the rivers, particularly in Bangladesh.

The mountains contain valuable minerals and gemstones, iron ore and coal. Many of the rivers have been dammed to produce hydroelectric power. The high mountains make it difficult to build roads and railways and many communities are very isolated.

◄ **Until the discovery of minerals the only people who ventured into the Rockies were hunters. Now national parks like this one in Wyoming are popular with many visitors.**

Rocky Mountains

The Rocky Mountains, often called the Rockies, are a series of mountain ranges running down the western side of North America from Alaska to Mexico. Some mountains have gentle slopes with rounded tops, but others are tall with jagged rocky peaks, many over 4,000 m (13,000 ft) above sea-level. Between the snow-capped peaks lie wide valleys, plateaux, lakes and rivers. In places there are hot springs, such as the geysers in Yellowstone National Park.

The Rockies started to form 190 million years ago, and are still rising slowly today. As the land has risen, the rivers have cut deep valleys and canyons in places. Many great rivers start in the Rockies. The mountains separate rivers flowing east, such as the Missouri, from those flowing west, like the Colorado River.

The Rocky Mountains are home to 5 million people. The mountains contain deposits of metals such as iron, silver, gold, lead and zinc, as well as uranium, phosphates and other salts. There is also coal, oil and natural gas. Many of the rivers have been dammed to produce hydro-electric power. It is generally the case that the higher up the mountains, the more plentiful the rainfall and snowfall.

The alpine meadows are full of wild flowers. Below the meadows, forests cover much of the mountain slopes, and cattle and horses graze on the grassy lower slopes. One of the most important sources of income is tourism, including skiing, fishing and walking in the many national parks.

Andes

The Andes are a large series of mountain ranges stretching 8,900 km (5,500 miles) down the whole west side of South America. There are many snow-capped peaks over 6,700 m (22,000 ft). In parts of the Andes, mountain ranges are separated by high plateaux, containing lakes such as Lake Titicaca. From the high, jagged peaks in the south, capped with permanent snowfields, glaciers push down to the Pacific Ocean. The formation of the Andes began about 80 million years ago and the mountains are still rising. There are many active volcanoes. The western slopes are desert or semi-desert in much of Peru and northern Chile, but forested further south.

Highest peak in the Rocky Mountains
Mount Elbert in Colorado 4,399 m

The Andes run through the countries of Venezuela, Colombia, Ecuador, Peru, Bolivia, Chile and Argentina.

Highest Andean peak
Mt Aconcagua in Argentina, 6,960 m

Volcanoes

Vulcano is an Italian island which has given its name to all other volcanoes.

Mauna Loa is the world's largest volcano. It reaches 4,170 m above sea-level, but its base is on the Pacific Ocean floor, 5,180 m below sea-level.

The volcanoes of the Andes, on the border of Chile and Argentina, are the highest in the world. Ojos del Salado (6,885 m) is the world's highest active volcano.

A volcano is a mountain or hill made of molten rock called lava which comes from deep beneath the Earth's surface. When a volcano erupts, lava and ash build up to make a cone. Some volcanoes give off clouds of ash and gas when they erupt. Others have streams of red-hot lava pouring down their side. Volcanoes can form on land or on the ocean floor. Some undersea volcanoes grow high enough to reach above sea-level and become islands.

The birth of a volcano

It is not often that anyone can see a new volcano appear and then grow. In Mexico, in 1943, some villagers were worried by earthquakes. Then a crack appeared across a cornfield and smoke gushed out. The crack widened, and ash and rocks were hurled high into the air. Soon, red-hot lava poured out. After a week, a volcano 150 m (500 ft) high stood where the cornfield had been, and the villagers had to leave. Mount Paricutín grew to 275 m (900 ft) in a year, and to 410 m (1,345 ft) after nine years.

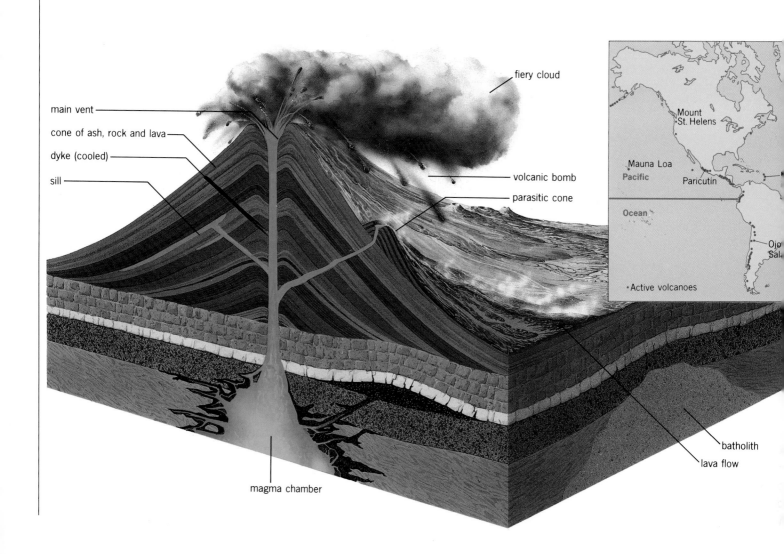

main vent
cone of ash, rock and lava
dyke (cooled)
sill
magma chamber
fiery cloud
volcanic bomb
parasitic cone
batholith
lava flow

Mount St. Helens
Mauna Loa
Pacific
Paricutin
Ocean
Ojos Sal.
Active volcanoes

How a volcano grows

The molten rock deep beneath the Earth's crust is called magma. It forces its way up through cracks and weak spots in the Earth's crust and spills out as lava. As magma rises, gases separate out from the molten rock. These gases may collect near the surface and cause a great explosion. On the island of Martinique, in the Caribbean, 20,000 people were killed by an explosion of hot gases and ash when Mont Pelée erupted in 1902.

When a volcano erupts, pieces of broken rock and ash are often thrown out with the lava. Large lumps are called 'volcanic bombs'. As the rock and ash cool, they make layers of solid rock. When Vesuvius, in southern Italy, erupted in the year AD 79, the town of Pompeii was completely buried under volcanic ash. Further round the Bay of Naples, Herculaneum was buried

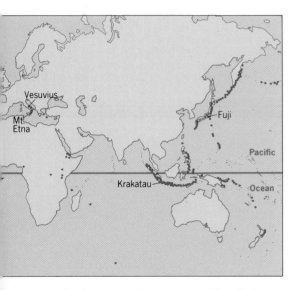

by mud which swept down the side of the volcano. Today, the remains of both towns have been dug out for all to see.

Dormant and extinct volcanoes

Volcanoes eventually die. A volcano that has not erupted for a long time is said to be dormant. There is always the danger that a dormant volcano may suddenly erupt. When people think a volcano has finally

died, then it is called extinct. Gradually, the volcano will be eroded. The softer rocks are worn away first, and in some places the only part left is the hard plug which filled the vent of the volcano.

There are about 700 active volcanoes in the world today, including some that are under the sea. The map shows that most of them are arranged like beads on a string. Notice how many volcanoes surround the Pacific Ocean. The areas with volcanoes are also the areas which suffer from earthquakes. This fact helped scientists come up with the theory of plate tectonics. They believe that the Earth's crust is broken into huge slabs, called 'plates'. Most of the world's volcanoes occur where plates meet, which is where magma can rise to the surface. A few, such as in Hawaii and the Canary Islands, occur above 'hot spots' where very hot magma seems able to pierce through a plate.

▲ Molten lava bursts out of Nyamulagira Volcano in Zaïre, Central Africa, and streams down the mountain.

The biggest explosion ever recorded was when Krakatau in Indonesia erupted in 1883. It was heard in India and Australia, 5,000 km (3,100 miles) away. It caused a giant wave which drowned 36,000 people.

Landforms

Landscapes are made up of landforms such as mountain peaks, lakes, volcanoes, waterfalls, cliffs and sand dunes. The science that studies landforms is called geomorphology. Geomorphologists are interested in the shape of landforms, the processes that make them the shape they are, and how their shape has changed through time.

Very few landscapes are flat. Most land slopes. Studying the angle and shape of slopes can give clues to the past. River valleys are one of the most common landforms. A large straight valley with steep sides and a flat floor may be a sign that it contained a glacier during the ice ages.

Landforms usually change slowly. It may take millions of years for rain, wind, frost and sunshine to demolish a mountain. The debris of broken rock or grains of sand is carried away by water, wind or the force of gravity to pile up somewhere else as new landforms.

At the coast, however, landforms can change relatively quickly. Waves erode the cliffs to form narrow beaches of pebbles and sand, and the wind blows the sand into lines of dunes that slowly move inland. The greatest changes occur during rare but fierce storms. Whole beaches may suddenly appear or disappear during a storm.

Landforms made by people

People make landforms too, especially with the help of powerful earth-moving machinery. Surface mines and gravel excavations leave behind huge holes, which may later fill with water.

Mining also produces heaps of waste. These artificial hills have sometimes been unstable. A wet spoil heap collapsed at Aberfan in Wales in 1966, burying a school and killing many children.

People have created dry land from marshes. The Dutch polders were built by constructing embankments or dikes around marshland and then pumping out the water. Good farmland is left, protected from flooding by the dike.

Other landforms made by people include straightened river channels and drainage ditches. At Hallsands in Devon the removal of shingle from the beach left the coast with no protection and the sea demolished a village.

On the Mississippi delta in the southern USA, walls have been built along the river channels to protect nearby towns and farms against flooding. But this also prevents the floods dropping sediment to build up the delta. Buildings that were once on the coast are left stranded on newly formed islands way out to sea.

The Bingham Canyon copper mine in Utah, USA is a hole 774 m (2,540 ft) deep. It is the world's largest excavation.

Erosion

▲ This weird Rocky Mountain landscape has been formed by erosion. The soft glacial silt has been worn away except where it is protected by caps of hard boulders.

The Mississippi River in the USA carries more than 500 million tonnes of eroded material to the sea yearly.

Although erosion is usually very slow it can sometimes be disastrous. In 1920 in Gansu Province, China, a sudden landslide killed 180,000 people.

You can check the speed of weathering of rock by examining gravestones in a local churchyard and noting their dates.

All around us, the Earth's surface is being worn away by water and by wind. This process is called erosion. Moving water includes mighty rivers and little streams, the sea and also ice which moves over the land as glaciers and ice-sheets. Water, ice and wind not only wear away the land, they also carry away eroded material and deposit it in other places, especially in seas and lakes.

Weathering

The action of snow and frost, sun and rain on rocks is called weathering. When rocks are exposed to the atmosphere, they are affected by the weather. Constant heating and cooling can split some rocks. When water in the rocks freezes it expands and cracks them. Rain water is a weak acid and can dissolve or change the chemicals in rocks. Weathering is also speeded up by plant roots and burrowing animals.

Rock pieces that have been broken up by weathering are moved away by water, ice and wind. When wind, water and ice are armed with pieces of rock, however small, they can erode even more powerfully.

The speed of erosion

Erosion is usually a slow process. But during storms, water and wind are much more powerful. They carry bigger fragments of rock and erode the land more quickly. A river in flood can erode the land faster in a few hours than it would normally do in years. The floodwater is armed with pieces of rock and fallen trees. It can roll great boulders along the river bed, and alter its course to cut new channels.

During storms, the great waves hurling water and stones at a cliff can be an awesome sight. After a storm, you can spot lots of signs of erosion along the coast. A sandstorm in the desert flings millions of hard sandgrains against the rocks, helping to erode them into strange shapes.

The rocks of the Earth's surface also affect the speed of erosion. Faults and folds make weaknesses that can be attacked. Soft rocks are eroded more quickly than hard ones. Hard rocks may make a waterfall along a river or headlands at the seaside.

New rocks from old

The pattern of erosion is part of a cycle. Eventually, the great mountain ranges will be worn down to become plains. The small particles of rocks carried by rivers and glaciers end up in the sea, building up sediments on the sea bed. After millions of years, they become sedimentary rocks such as sandstone and clay. Eventually, those rocks may be pushed up to make new land and new mountain ranges. This land in its turn will be worn down by water, ice and wind.

Glaciers

A glacier is a moving sheet of ice. Glaciers usually form when enough snow builds up in ice layer on the land. They can be 100 m (330 ft) high. If it is on a slope, the great weight of the ice causes the whole sheet to move downhill. This is because the layers of ice at the bottom of the pile where it rests on the ground become softened and slippery, and the glacier begins to move like a very slow river.

The glacier usually begins high in a mountainous region, and it moves downhill in fits and starts. The ice may move smoothly for a while, until it comes up against an obstruction. This may be a bend in the valley, or a mound of broken rocks which have been pushed ahead by the glacier itself. Glaciers move at different speeds, usually between 1 cm (0·4 in) and 1 m (3 ft) per day. The middle part of the glacier moves faster, and the edges, which rub against the sides of the valleys, may be much slower. Glaciers may move downhill and pass right into the sea. When they enter the sea, large pieces break away and float off as jagged icebergs.

How glaciers change the landscape

Glaciers that flow in valleys make their valleys wider and deeper. The bottom of a glacier carves out a U-shape. The ice simply cuts away the soil and boulders, and pushes them ahead. At the front edge (the 'snout'),

rocks and soil may be dumped in untidy heaps called moraines. The melt-water in summer carves out narrow streams that run downhill in front of the glacier. These streams can carry waste material worn away by the ice for many kilometres.

Evidence of past glaciers can be found over a very large area. Most of the British Isles and northern Europe, for example, lay under ice during the last ice age. Here, most of the upland valleys were widened by glaciers while the lowlands were covered by sands, clay and boulders.

Glaciers can be seen today in the European Alps and other high mountain ranges, in Alaska and northern Canada, New Zealand, Greenland and Antarctica.

Over 10 per cent of the Earth's land surface is permanently covered by ice.

The largest glacier in the world is 514 km (320 miles) long. It is the Lambert Glacier in Antarctica.

The fastest glacier in the world is the Quarayac Glacier in Greenland, flowing at a rate of 20 m per day.

valley carved into a U-shape by the glacier

rock debris

cracks or crevasses

stream of melt-water

glacier snout

moraine

Rivers

► The course of a river from its source in the mountains to its estuary at the sea.

waterfall

sea

meanders

oxbow lake

delta

tributary

Most river basins are worn down by about 1 m every 30,000 years.

A river is formed when water flows naturally between clearly defined banks. The water comes from rain or snow. When rain falls or snow melts, some of the water runs off the land down the steepest slope, forming trickles of water in folds of the land. These trickles eventually merge together to form streams, which join up to form rivers. The streams which join the main river are called tributaries. Some of the rain-water also sinks into the ground, and seeps down through the rocks until it meets a layer of rock which cannot hold any more water. Then the water runs out at the surface to form a spring.

A river gets bigger and bigger as it flows towards the sea, because more and more tributaries join it. The area of land which supplies a river with water is called its drainage basin.

Rivers wear away rocks

Rivers cut into the land and create valleys and gorges. Rushing water has tremendous force. A cubic metre of water weighs a tonne. Water can split rocks just by pounding them. But more important is the load of sediment (stones and sand) the river carries. Rocks and soil are swept along by fast-flowing water, scouring the river bed and banks. Large boulders are bounced along the river bed, scouring out a deeper and deeper channel.

The rate at which the water wears away the land depends partly on how hard the rock is, and partly on the slope of the river. The steeper it is, the greater its power to erode (wear away). Where the land is rising or the sea-level is falling, rivers can cut down through the rocks very fast. The mountains of the Grand Canyon in the United States were rising as the Colorado River cut down through it. Today, the river has cut a gorge 1·5 km (1 mile) deep.

The faster a river flows, the larger the rocks and the greater the load of sediment it can carry. When the river's flow is slowed down, it drops some of the sediment it is carrying, the largest pebbles first, then the sand, and finally the fine silt. This happens when the river enters the still waters of a lake or the sea, or when the valley floor becomes less steep as it leaves the mountains.

The river's upper course

Near its source the river is well above sea-level and is flowing very fast, so it has its greatest cutting power. The water sweeps along boulders and pebbles. The boulders grind against each other, gradually breaking down into smaller pieces of gravel, sand and mud. The river is still small, and quite shallow. Its bed is full of boulders.

The river's middle course

Here, the river is not flowing so fast. It contains more water, so its bed is wider, and is lined with sand, small pebbles and water weeds. The river is not powerful enough to rush over large obstacles, so it flows round them, and its course winds among the hills. Where the water swings around a bend, the water on the outside of the bend has to flow further than the water on the inside, so it flows faster. It cuts away the bank on the outside of the bend, widening the valley.

The river estuary

As the river nears the sea, it becomes wide and sluggish, making huge curves (called meanders) around the slightest obstacles. As it spreads out and slows down, it sheds its load of sediment. It wears away tiny cliffs on the outside of the meanders and deposits little beaches of sand on the inside of the bends. When the river floods, it flows over its banks, spreading mud and sand over the surrounding land. As it enters the sea, it builds out a fan-shaped delta of mud.

A river's life has three stages from its beginnings in high ground to its joining with the sea.

◄ The fast-flowing upper course of a river.

▼ In its middle course a river flows more smoothly, often through a wide valley.

◄ The sluggish final stage in a river's life as it meanders towards the sea. The wide mouth of the river where it joins the sea is its estuary.

Seasonal changes

Many rivers have very different flows in summer and winter. In cold regions, the upper part of the river may be frozen in winter, so flow decreases. Melting snow and ice may cause spring floods. Rainfall may be seasonal, so there are annual floods. In arid lands, some rivers exist only for a brief period after heavy rain, when water rushes off the bare, baked soil, carrying huge boulders and cutting deep gorges called wadis. Other desert rivers vanish long before they reach the sea, as the water simply evaporates or sinks into the sand.

Rivers carry about 8 thousand million tonnes of sediment into the oceans every year.

Nile

▲ **Women washing their clothes in the River Nile near Luxor, Egypt. Away from the river the vegetation soon gives way to desert.**

The Nile drains one tenth of the continent of Africa.

The Sudd swamp in Sudan is the probably the largest wetland in the world. It covers 129,534 sq km (50,000 sq miles).

The River Nile is the longest river in the world, 6,695 km (4,160 miles) from source to sea. It drains one tenth of the continent of Africa. The Nile flows out of Lake Victoria and through a series of spectacular gorges before spreading out across the great papyrus swamp called the Sudd in Sudan. It then becomes the White Nile, and is joined by the Blue Nile and the Atbara River, which both start in the Ethiopian highlands. The Nile then wanders across its flood-plain to Cairo, where it forms a huge delta, 250 km (155 miles) wide, before reaching the Mediterranean Sea. Soil carried down by the river during the annual floods keeps the Nile delta fertile.

In 1970, the Aswan High Dam was completed, damming the Nile to form Lake Nasser. The dam regulates the flood waters, generates hydroelectric power, and irrigates thousands of acres of formerly unproductive land. But it has reduced the supply of minerals to the Nile delta, and so the farmland is less productive and coastal fisheries have been harmed.

The Nile is navigable for most of its course through Sudan and Egypt, except where the water is very low and there are rapids, the Cataracts.

The papyrus beds of the Sudd swamp are home to a rich variety of wildlife, including the shoebill heron and sacred ibis, crocodiles and hippos, and hundreds of thousands of antelopes.

The Sudd is threatened by the Jonglei Canal, which will take water from the Nile, and link up important towns. At present, work on the canal is held up by civil war. If it is ever complete, the Sudd may dry up. Local people will lose a vital source of water and grazing land, and the rich wildlife will disappear.

Mississippi River

The Mississippi River together with its main tributary, the Missouri, is the largest river system in North America. The Mississippi-Missouri flows a total 6,019 km (3,740 miles) and drains an area of 3,221,000 sq km (1,244,000 sq miles). Every day it discharges 1,600 million tonnes of water into the Gulf of Mexico.

The Mississippi rises in Lake Itasca, west of the Great Lakes, and is itself 3,779 km (2,348 miles) long. As well as the Missouri, which rises in the Rocky Mountains, other large tributaries are the Ohio River, which starts in the Appalachian mountains, and the Arkansas River. As it flows towards the sea, the Mississippi grows from a clear stream winding its way through lakes and marshes, to a huge muddy river over 2½ km (1½ miles) wide, which wanders in huge sweeping curves across its flood-plain. The huge amount of sediment carried by the river has built up a wide delta pushing out into the Gulf of Mexico near New Orleans, and the river splits into hundreds of tiny rivulets as it meanders over the delta.

In the first half of the 19th century the Mississippi was very important to the economy of the South and Middle West of the USA. Steamboats carried cargo and passengers, and showboats brought theatre to the waterfront towns. It was a colourful and glamorous period. But after the American Civil War and the coming of the railways the Mississippi never quite regained its former glory.

It is still an important highway for transporting cargoes of iron, steel, coal, petroleum, chemicals and other raw materials and industrial products. In places, the river has been straightened and its banks have been raised to improve the passage of river traffic and hold back flood water.

The huge Mississippi flood-plain is over 125 miles wide in places, and covers about 77,720 sq km (30,000 sq miles). The fertile silt brought down by the floods has given rise to valuable farmland.

Amazon River

The Amazon River is 6,440 km (4,000 miles) long, the second-longest river in the world. It contains more water than any other river in the world - about 25 per cent of all the water that runs off the Earth's surface.

The Amazon rises high up in the Andes mountains in Peru, and is fed by about 15,000 tributaries on its way to the sea on the coast of Brazil. The land slopes gently, and the Amazon wanders in great curves called meanders, frequently changing its course as the river bed becomes silted up, leaving behind little horseshoe-shaped lakes and swampy areas rich in birds and other wildlife.

The river contains over 2,000 different kinds of fish, including piranhas, catfish, electric eels and the giant arapaima. The nutrient-rich silt deposited by the river supports vast tropical forests which line its banks, home to millions of different animals and plants.

Thousands of small villages along its banks live off the fish and small plots of land which they clear in the surrounding rainforest. Large ships can travel far inland to cities such as Manaus in Brazil.

Every second the Amazon River carries an average of 120,000 cubic metres (156,960 cubic yards) of water into the Atlantic Ocean.

The Amazon River basin (the land drained by the Amazon) is the largest in the world, covering about 7,045,000 sq km (2,720,000 sq miles).

The Mississippi delta covers 38,860 sq km (15,000 sq miles).

▼ The city of New Orleans lies on the Mississippi delta. It is an important port for vessels carrying goods between the American south and the Gulf of Mexico.

Valleys

▲ **The Colorado River has cut steeply into the surrounding mountains, forming the spectacular Grand Canyon in the USA. There, the mountains were rising as the river was cutting down. Because of low rainfall the sides of the gorge have weathered little.**

There are valleys on the sea-bed as well as on land. The deepest underwater canyon is south of Western Australia. It is 1,800 m deep and 32 km (20 miles) wide.

Valleys are formed by the action of rivers or glaciers, wearing away the rocks. As the water or ice flows down from the mountain tops it carves out valleys, leaving ridges of rock in between and giving the mountains their shape.

River valleys

Rivers flowing down steep gradients (slopes) have great cutting power. Rocks and boulders roll down the valley sides into the river and are bounced along its bed, cutting deep into the rocks. River valleys high in the mountains are steep-sided. If you see them in cross-section, they look V-shaped.

Further down the river course, the gradient of the river bed is less, the river flows more slowly and cuts less deeply. It develops a more winding course, flowing around obstacles and cutting a wider valley, like a shallow U. The river flood-plain has a very slight gradient and may appear almost flat. Some river flood-plains are hundreds of miles wide.

Glaciated valleys

Glaciers can erode valleys much more powerfully than rivers. The ice may be hundreds of feet thick, so a great weight presses down on the valley floor. Rocks and boulders become stuck in the ice on the bottom of the glacier. The ice pushes them along like a giant piece of sandpaper. Glaciated valleys are U-shaped in cross-section, with very steep sides and flat bottoms. After the ice has melted, the rivers that flow down these valleys look much too small for them.

Where tributary glaciers enter the main valley they form 'hanging valleys'. Because these glaciers are smaller, their valleys are not cut so deep. They are seen high up the main valley sides. Picturesque waterfalls often cascade from these hanging valleys after the ice has retreated.

Gorges

Where the river follows a line of weakness in the rocks, such as a fault, it may cut a very deep valley with almost vertical sides, called a gorge. Gorges also form in desert areas where rainfall occurs very rarely in the form of heavy thunderstorms. The soil has been baked hard by the Sun and very little water can soak in, so huge volumes of water run off the surrounding land, forming 'flash floods' so powerful that they carry large boulders with them and cut deep gorges called wadis.

Some of the most spectacular gorges are the canyons of the United States, including the Grand Canyon, 1·5 km (1 mile) deep.

Dry and drowned valleys

In some parts of the world, especially where there is limestone or chalk, there are valleys which contain no water. These dry valleys were formed at a time when the climate was much wetter, as at the end of the last ice age. Their rocks are very porous, and water quickly sinks down into them. When there was more rainfall, the water level in the rocks remained above the valley floor, so the rivers flowed. Today, the water is much deeper and only underground rivers occur.

Where coastlines are sinking or sea-levels rising, valleys may be drowned, forming long inlets of the sea called rias. Rias are found on the south-west coast of England. Where deep glaciated valleys are drowned, they form fjords with very steep sides and extremely deep water, as in Norway.

Rift valleys

Some of the widest valleys in the world are the rift valleys. These were formed when huge blocks of rock moved relative to each other: either the blocks on either side of the valley were raised up to form mountains, or the block in the middle dropped down to form the valley floor. The great Rift Valley of Africa and the Great Glen of Scotland were formed in this way.

Valley settlements

In mountain regions the valleys are the main areas where people live. This is because the climate is milder and the soils are thicker. The valleys provide routes for roads and railways, and larger rivers can be used for transport, too. Large settlements often occur where one or more valleys meet, encouraging trading.

▼ The slow progress of this river in Germany has allowed time for the valley sides to weather. The result is a wide valley with gently sloping sides.

▶ **Farmland in the Hunza valley in the mountains of Pakistan. The floor of the valley is terraced and a variety of crops grown, including wheat, rice and vegetables. The valley is famous for its apricots.**

Where springs emerge on the valley slopes there may be a line of villages which date back to times when spring water was the main source of water for drinking. In valleys with fair-sized rivers water power may be used to generate electricity for industry.

Communications between valleys may be difficult in mountain country. Roads and railways have to climb winding mountain passes, or pass through expensive tunnels. The hills may also interfere with radio and television reception.

Life in valleys

A valley floor is lower than the hilltops, so the climate here is warmer. The valley sides provide shelter from wind; often the mountains themselves receive much of the rainfall and snowfall, leaving the valleys short of rain. South-facing valley slopes get more light and are usually warmer than north-facing slopes, and often have different vegetation.

In some valleys there are special mountain winds which blow down the valley bringing warm or cold air at certain times of year. When frost and fog form, the cold air sinks, and valleys often have more frost and fog than the mountains above.

In high mountain valleys the soils are usually thin and not very fertile. Sheep farms are more common than crops. On very steep slopes, farmers may build terraces to prevent the water and soil rushing away down the slope. Where there is no farming the slopes may be wooded. The trees help to anchor the soil and prevent it being washed away in heavy storms. They also help to prevent avalanches of snow or mud crashing into the valley and villages below. There are often trees and shrubs near the river, benefiting from the extra moisture.

In lowland valleys flooding caused by seasonal rainfall or melting snow may be a problem. But farmers need the rich silt that is left behind by the floods. In valleys near the river mouth the soil is even thicker. Good crops can be grown in the fertile sediments dropped by the river.

Where forests have been cut down in the hills, flooding may be much worse. With no trees to soak up the water, more of it rushes straight into the rivers, together with the soil.

Mountain migrations
In many mountain areas, such as the Alps and Himalayas, cattle are moved up to high alpine pastures in spring as the snow melts, and brought back to the shelter of the valley floor in winter to be fed on hay.

Lakes

Lakes are areas of water surrounded by land. They occur where water collects in hollows in the Earth's surface, or behind natural or man-made barriers.

Lakes don't last forever. The water may cut through the barrier, so the lake drains away. Sooner or later most lakes fill up with sand and mud. As a river enters a lake, the water flows slower and drops its load of sediment. Plants grow in the sediment, trapping more sand and mud.

Lakes also disappear if more water flows out of them or evaporates than the rivers bring in. When a desert lake evaporates, the dissolved salts and sediments are left behind and gradually fill up the lake, which becomes very salty. The Caspian Sea is like this. It has shrunk drastically as more and more irrigation water has been taken from the Volga and Ural rivers which feed it.

▲ Probably the most famous of all lakes in volcanoes is the Crater Lake in Oregon, USA, which is 9 km (6 miles) across.

HOW LAKES ARE FORMED

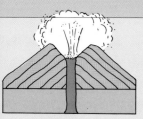

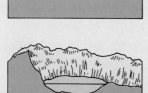
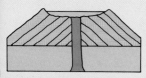

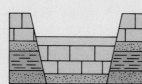

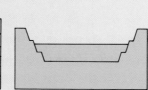

A crater lake is one which lies in the natural hollow of an old volcano. The Eifel district of north-west Germany has hundreds of lakes lying in extinct craters. One of the rarest crater lakes is Lake Bosumtwi in the Ashanti Crater in Ghana. The crater was probably made by a meteorite.

Glacial lakes form where ice-sheets and glaciers have left the ground very uneven. They scraped and hollowed out hard rock or dumped sand, gravel and clay in uneven layers. Finland is a country of such lakes. Northern Canada and north-west England have similar lake districts.

Rift valley lakes are long thin lakes such as Lake Malawi, Lake Turkana and Lake Tanganyika in East Africa, the Sea of Galilee in Israel and the Dead Sea between Israel and Jordan. When the Earth's crust slipped down between long lines of faults, the water filled part of the valley floor.

Artificial dams have created lakes. People have built earth, stone and huge concrete dams to hold back rivers for water supply, irrigation or hydro-electric power. Lakes may form in disused gravel pits and mines. Often they are used for leisure and water sports, or to attract wild birds.

Largest lake
Caspian Sea, 371,000 sq km (143,205 sq miles)
Largest freshwater lake
Lake Superior, 83,270 sq km (32,140 sq miles) (border of Canada and USA)
Deepest lake
Lake Baykal, 1,741 m deep (in Siberia, Russia)
Highest navigable lake
Lake Titicaca, 3,811 m above sea-level (in the Andes of Peru and Bolivia)
Largest temporary lake
Australia's Lake Eyre, a desert lake 9,300 sq km (3,600 sq miles) in area. It disappears completely after a few dry years.

Oceans and seas

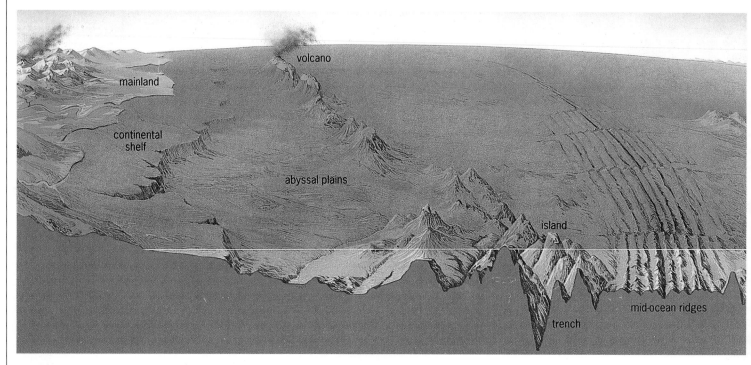

mainland

volcano

continental
shelf

abyssal plains

island

trench

mid-ocean ridges

▲ **A cross-section of the ocean floor.**

The five major oceans are the Arctic, Atlantic, Pacific, Indian and Southern Oceans. They are connected to each other by open water. Water slowly circulates between them in currents at the surface and deeper down. The oceans contain about 1,370 million cubic km (330 million cubic miles) of water altogether. The average depth of this water is 4,000 m (13,000 ft), but in some ocean trenches it may be 11,000 m (36,000 ft) deep.

The ocean basins

The ocean floor has a landscape of its own. Much of the deep sea-bed is a flat plain. But in places, mountains rise thousands of metres from the sea-bed, sometimes pushing through the sea's surface as islands. Many of these are active or extinct volcanoes. Running down the centre of the ocean floor in several of the oceans is a ridge of mountains which is continually being built up by outpourings of lava. As the rock is forced outwards from the ridge by the new lava, the ocean floor spreads until it reaches the boundaries of the continents. At the edge of each continent is a shallow

shelf which slopes gently down to about 200 m (650 ft), then dips steeply down, in some cases to a deep trench which marks the point where the ocean floor is being forced under the continent.

The ocean floor

Much of the ocean floor is covered in sand or mud brought in by rivers. In places, hot springs bubble up, depositing sulphur and other minerals. Millions of microscopic plants and animals live in the surface waters. When they die, their glassy or chalky shells sink down to the bottom to form a sediment. Here, the pressure of the water above and the pressure of other sediment layers slowly turn the sediments into rock. Future upheavals of the Earth's crust may one day fold these rocks into new mountain ranges and new land.

Moving water

The water in the oceans is constantly moving, driven by winds, waves, tides and currents. It may be moving in different directions and at different speeds at different epths. Where the wind blows from the ame direction for most of the year, it is ble to move large volumes of water, orming surface currents. But the spinning f the Earth on its axis makes these urrents turn to the right in the northern emisphere, and to the left in the southern emisphere. So the surface currents move n giant circles called gyres.

f you mix oil and water, the oil floats on he top because it is less dense than the vater. Warm water is less dense than cold vater, and salty water is denser than fresh vater. In the oceans, cold or salty water inks, and this sets up deep currents.

Cold and warm currents

n the tropics, the warm surface waters are oushed into two great west-flowing urrents by the north-east and south-east rade winds. Between them, the equatorial ounter current flows in the opposite direc- ion to compensate. Where these currents each the continents, the rotation of the arth forces them into clockwise circles in he northern hemisphere, and anticlock- vise circles in the southern hemisphere.

Nearer the poles, these circular currents meet cold water flowing from the melting ce, and return to the equator as cold urrents. Where cold water wells up from he deep ocean, it brings nutrients which

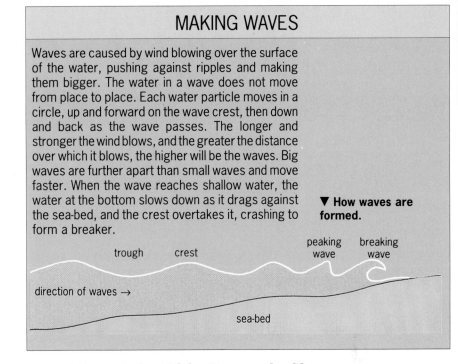

MAKING WAVES

Waves are caused by wind blowing over the surface of the water, pushing against ripples and making them bigger. The water in a wave does not move from place to place. Each water particle moves in a circle, up and forward on the wave crest, then down and back as the wave passes. The longer and stronger the wind blows, and the greater the distance over which it blows, the higher will be the waves. Big waves are further apart than small waves and move faster. When the wave reaches shallow water, the water at the bottom slows down as it drags against the sea-bed, and the crest overtakes it, crashing to form a breaker.

▼ **How waves are formed.**

trough   crest   peaking wave   breaking wave

direction of waves →

sea-bed

support large stocks of fish. Warm and cold currents also affect the climate of coastal countries.

Frozen seas

Near the poles, parts of the Arctic and Southern Oceans form permanently frozen ice shelves stretching out from the coast. In slightly warmer parts, the sea freezes in winter, forming pack ice up to 2 m (7 ft) thick. In summer it thaws, and the pack ice breaks up into flat-topped icebergs.

Deep cold currents are set up by cold salt water sinking at the poles and spreading over the ocean floor. These currents can carry water all the way from the North Pole to the Pacific Ocean.
A cold water current 5,000 m deep circles around Antarctica, moving an amazing 190 million cubic metres of water a second.

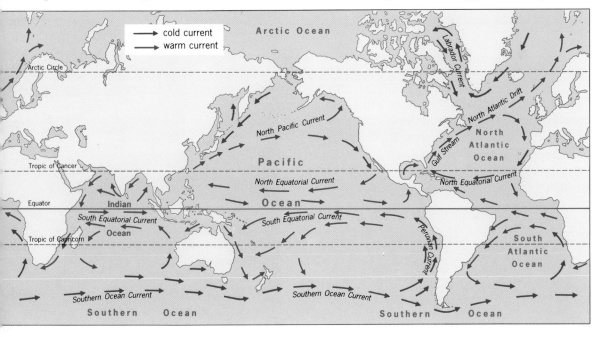

◄ **Oceans cover more than 360 million sq km (140 million sq miles), over 70 per cent of the Earth's surface.**

Icebergs and ice-caps

polar regions. Cold currents carry icebergs great distances. When the cold Labrador Current continues further south than usual, it carries Arctic icebergs into the busy North Atlantic shipping routes. The International Ice Patrol looks out for drifting bergs. Ships can be warned, and sometimes the bergs themselves can be towed away by powerful tugs.

Ice-caps and ice-sheets

Ice-caps and ice-sheets are large areas of ice and snow which permanently cover the land. Ice-sheets are larger than ice-caps. Today, the main ice-sheets of the world are found in Greenland and Antarctica.

Here, temperatures stay at or below freezing all the year round. Snow accumulates

▲ **Arctic icebergs are typically tall and uneven. They may drift hundreds of kilometres and are often shrouded in fog.**

▶ **Antarctic icebergs tend to be flatter-topped and larger than those in the Arctic.**

Largest iceberg
More than 31,000 sq km (335 km long and 97 km wide, bigger than Belgium), seen in the South Pacific Ocean in November 1956
Tallest iceberg
167 m high, seen off west Greenland in 1958

Icebergs are huge lumps of ice which have broken away from ice-sheets and glaciers and are floating in the sea. Only about a ninth of an iceberg shows above the surface. The part that is hidden under the water may be wider than the part that shows. This is a great danger to shipping. The liner *Titanic* sank after hitting an iceberg in the North Atlantic in 1912.

The largest icebergs break away from the edge of Antarctica, such as from the Ross Ice Shelf. This area of floating ice is as large as France. The front of this shelf is 650 km (400 miles) long, with ice-cliffs 50 m (160 ft) high.

In the Arctic, icebergs break off from the Greenland ice-sheet. They are not as large as the Antarctic ones, but are often taller. Glaciers reach the sea around Greenland, and as the ice begins to float, huge lumps break off. This is called 'calving'. Icebergs begin to melt as they drift away from the

throughout the long winter, but very little melts in the short summer. As the snow builds up, it turns to ice under its own weight. This great weight makes the ice spread out and flow downhill. It moves down valleys to form glaciers. If the ice reaches the sea, it may spread out to form an ice-shelf, or it may break up into icebergs.

The weight also weighs down the land on which it rests. If all the ice on Greenland and Antarctica melted, the land would rise. But if all the world's ice melted, the sea-level everywhere would rise at least 65 m (210 ft). Huge areas of lowland would be drowned, and the shape of the continents would change.

Ice ages

The present-day ice-caps and ice-sheets are all that remain of much larger areas of ice which spread out over much of Europe, Russia and North America in the ice ages.

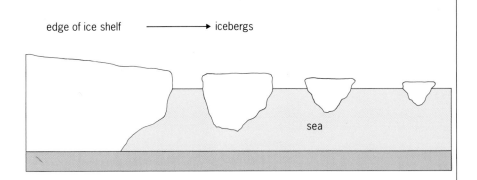

As the ice-sheets moved over the land, they changed it. Huge areas of rock were scraped bare and the fragments of rock scraped off were dumped at the edge of the ice-sheet. These deposits are called moraines. During periods of warmer weather, the ice melted and the edge of the ice-sheets retreated. Moraines were dumped on the land and spread by water flowing from the melting ice. The hills and hollows left by the ice created many lakes and new river courses. Some rivers were dammed by moraines, while others followed new courses which had been cut by the ice.

▲ Icebergs form when huge lumps of ice break off from the edge of the ice-sheet when it reaches the sea.

Rocks

There are three main types of rocks. Igneous rocks are formed at very high temperatures from molten liquids, either deep within the Earth, or at the surface, from volcanoes, for example. Sedimentary rocks are formed from sand, mud, or limy mud, laid down on land or in ancient rivers, lakes or seas. These form sandstone, mudstone and limestone. Metamorphic rocks are formed from either sedimentary or igneous rocks that have been buried and heated up or put under great stress.

Igneous rocks

Igneous rocks are generally very hard, because they are formed from a molten mass of rock material called magma. The magma, which can be seen on the surface of the Earth as lava, consists of complex mixtures of chemicals. As the magma moves through cracks in the Earth's surface, it cools around the edges. It may cool completely deep within the Earth, or at the surface.

The kind of igneous rock formed depends on the composition of the magma and the rate at which it cools. On cooling, the chemicals in the magma form crystals. These crystals, like grains of salt or sugar, have particular shapes depending on their chemical composition, and the longer it takes them to cool, the larger they become. The solid crystalline forms are called minerals.

Igneous rocks formed deep in the Earth's crust are called plutonic rocks. A common type is granite, which contains many large crystals of the common mineral quartz. Igneous rocks formed nearer the surface in dikes or sills include dolerite, which contains smaller crystals of other minerals, but hardly any quartz. Igneous rocks which are formed at the surface are called volcanic rocks. They include basalt, a common rock type which is more fine-grained than dolerite because it cooled faster at the surface.

dolerite

granite

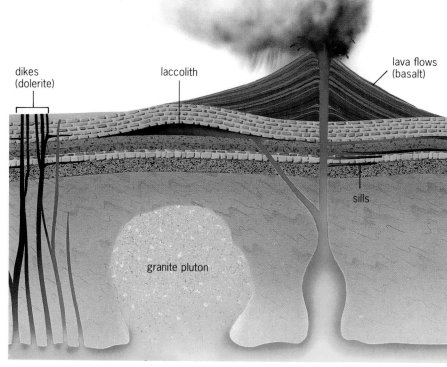

dikes (dolerite)

laccolith

lava flows (basalt)

granite pluton

sills

basalt

◄ Only a small proportion of the molten lava which forces its way up through the Earth's crust eventually solidifies on the surface. Most cools within the crust, squeezed between existing rocks. The strange shapes that cool underground are often exposed millions of years later, as weathering wears down the surface of the Earth.

Sedimentary rocks

Sandstones, mudstones and limestones were once sands and muds on the bottom of rivers, lakes and seas. These sediments form part of the great cycle of erosion and deposition that is going on all of the time. The rocks on land are eroded by the action of rain, streams, wind, ice and plant roots. Small rock fragments in the form of sand or mud are blown away or washed down into rivers or lakes where they may sink to the bottom. Over the years, millions of sand grains may be deposited in one place, and thick layers build up.

The largest areas of deposition are in the sea. Sediment is brought in from all the great rivers, and it is also eroded from coastlines by the sea itself. These grains of sand and mud may be deposited in shallow coastal waters, or they may be swept well out into the oceans. Over millions of years, the sediment may build up to be many hundreds of metres thick.

The land here is made from sediments deposited by the river during the last 5,000 years.

Sediment deposited by the river when it reaches the sea.

From sediment to rock

For the sediment to form rock, two main changes take place. First of all, water is lost. As the layers of sediment build up, the deeper sediment is pressed down by the weight of sand or mud above. The grains press closer together, and water moves up into the lake or sea above.

The second stage is the formation of a kind of cement. After many years, sometimes thousands of years, the remaining water around the grains contains solutions of minerals. Further loss of water causes these mineral solutions to form crystals in the spaces between the rock grains.

▲ A river carries sediment to the sea. Sediment builds up at the mouth of the river, spreading wider and becoming thicker. Eventually the sediment hardens to form sedimentary rock.

▲ Millstone grit is a coarse sandstone. The grains of sand can easily be seen without a magnifying glass. This rock was once used to make millstones for grinding corn and has been much used as a building stone.

▲ Chalk is a soft white porous rock. It is a different chemical from that of blackboard chalk but can look very much like it. It is made from the remains of chalky algae and often contains knobbly lumps of flint.

▲ Shelly limestone is made from the piled-up skeletons of sea animals. Usually the shells were broken up by tides at the time the limestone was first deposited, but sometimes these fossil shells are perfectly preserved.

Metamorphic rocks

The word metamorphic means 'later, or changed, form'. These are rocks which have been altered either by heat, or by heat and pressure together.

When magma forces its way up in the Earth's crust, it heats the surrounding rock for a distance of several metres. If it is passing through sandstone, this may be baked into hard quartzite. Limestone may be baked into marble. This type of metamorphism can affect only small amounts of rock close to the passage of the magma.

Large-scale metamorphism can take place when mountains are forming. For example, when major plates of the Earth's crust collide, there are great pressures, and ocean-floor sediments like mudstones and sandstones may be altered over wide areas to form slates and schists. Both pressure and heat are involved.

Changing crystal direction

When a metamorphic rock forms, the pressure forces all the crystals in a rock to line up. In mudstones and sandstones, all the mineral crystals lie in a random arrangement. When great forces act on these rocks, the grains line up at right angles to the direction of the pressure.

New minerals

New minerals are often created by metamorphism. The great heat can melt the original rock into a liquid, and impurities may then come together and form new minerals such as tourmaline or garnets on cooling. A very unusual example of this is the formation of diamonds in coal layers which have been heated by igneous rocks.

► Slate splits into thin sheets. However, cutting roof slates by hand is skilled work. Fewer stonemasons have this skill today as roofs are more usually covered with clay tiles.

▲ Gneiss is a coarse-grained rock with irregular bands of different colours. Gneisses often glitter because they contain the mineral mica. They were formed from rocks changed by great heat and pressure.

▲ Slate is made from clay rocks transformed by heat and pressure. The most distinctive feature of slate is its fissility – the ability to split cleanly into thin sheets. It is best known as a material for roofing.

▲ Marble is made from limestones changed by great heat and pressure. The colours and patterns in marble depend on the type of limestone from which it was formed. Pure white limestone formed the valuable Carrara marble from Italy, which sculptors, such as Michelangelo, used.

Soil

oil is formed when rocks are slowly broken
own by weathering (the actions of wind,
ain and other weather changes). Plants
ke root among the rock particles. The
oots help to bind the particles together, and
rotect them from rain and wind. When the
lants die, they decay and produce a dark
ticky substance called humus. The humus
ticks the soil particles together and absorbs
ater.

The soil environment

oil is made up of a mixture of rock particles
f various sizes, with air spaces between
hem. The particles are coated with humus
nd a thin film of water. The larger the soil
articles, the bigger the air spaces between
hem and the faster water drains out of the
oil. The air spaces are important for plants
ecause their roots need oxygen to breathe.
he humus supplies minerals to the plants
s it decays.

Soil profiles

ifferent kinds of parent (original) rock and
ifferent climates produce different kinds of
oils. You can see this by looking at soil
rofiles. A soil profile is a sample taken from
he surface down through the soil. Each
rofile is divided into a series of layers called
orizons.

) horizon, the surface layer, contains
nany plant roots and soil animals. It is rich
n dark-coloured humus.

\ horizon still has a lot of humus, but is a
aler, greyish colour because many of the
ninerals have been washed out by rain-
vater, a process called leaching.

3 horizon contains much less humus, but
ome of the minerals washed out of the A
orizon are deposited here. If the soil is not
oo wet, any iron left here will oxidize, pro-
lucing a yellow or reddish brown colour.

: horizon is where weathering is taking
lace, and the parent rock is broken down.

k horizon is the parent rock.

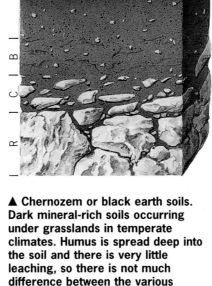

▲ Chernozem or black earth soils.
Dark mineral-rich soils occurring
under grasslands in temperate
climates. Humus is spread deep into
the soil and there is very little
leaching, so there is not much
difference between the various
layers.

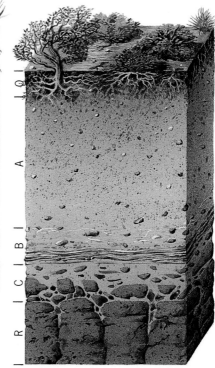

▲ Podzols.
Poor soils occurring in regions of
heavy rainfall or acidic rocks, where
minerals are rapidly leached from
the upper layers. The minerals are
often deposited lower down as a
hard iron-rich layer called a pan.

Soil animals

One square metre of fertile soil contains
over 1,000 million animals. Many of them
are too tiny to see with the naked eye.
Fungi and bacteria break down plant and
animal remains, releasing minerals which
are then absorbed by plant roots. Earthworms tunnel through the soil, letting air
in, helping water to drain through, and
mixing the different layers. Earthworms can
move 10 tonnes of soil per hectare per
year. Ants, beetles, centipedes, millipedes
and spiders hunt in the soil, and larger
animals like foxes, rabbits and mice make
their burrows there.

In some parts of the world
wind-blown dust
accumulates to form a soil
called loess. In parts of
China the loess is 300 m
(1,000 ft) thick.

One square metre of soil
contains over 1,000
million animals, many of
them too small to see.

Fossils

HOW DO FOSSILS FORM?

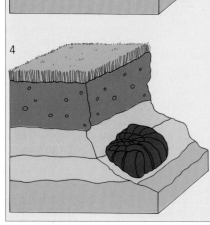

The commonest sorts of fossils show us the hard parts of the ancient plant or animal. The hard parts include the materials making up shells and bones, which do not rot away. Fossils are formed in several stages, from the body of a dead animal, or from a whole plant, or from parts of a plant.

Imagine many shellfish living on the sea-bed in shallow water. Some of them may burrow into the sand on the sea-bed, while others move about on top of the sand, and some may be fixed to boulders. A great storm could whip several metres of sand over them all, and kill them before they could escape.

Within a few weeks, all of the soft insides of the shellfish will rot away, or be eaten by creatures that burrow through the sand. Over a few thousands of years, more layers of sand may be dumped on top, and water may be forced upwards by the weight. The sand settles, and it may eventually form a rock called sandstone, in which all the grains are firmly stuck together.

The shells are deep inside the sandstone. Mineral solutions may pass through the sandstone, and some of these harden the rock and the fossils. The space inside the shells, where the soft parts were, may also fill up with crystal deposits from these mineral solutions, so that the shells are literally turned to stone or rock.

After millions of years, the coastline changes and the sandstone is on dry land. Then the landscape is cut down by wind and rain, or a stone quarry is dug. The ancient shells are found as fossil shells.

◄ How fossils are made

1 300 million years ago. An ammonite swims in the sea.

2 The ammonite dies. Its shell sinks to the sea floor and the fleshy parts rot.

3 The shell is buried by sand and mud. It is gradually replaced by new minerals.

4 Today. The fossil ammonite is exposed on a cliff face.

Fossils are the remains of plants and animals that once lived on the Earth. Often they show only a part of the plant or animal, usually a hard part, such as a shell or a bone.

Common fossils

The commonest fossils are ancient sea shells. This is because most of the rocks that contain fossils were laid down in the sea, and because shellfish have always been very common. It is easy to find other sea creatures. For example, the Ordovician and Silurian mudstones of many areas contain fossil graptolites (animals which floated near the sea surface), while limestones of the same age contain corals and brachiopods (a particular group of shellfish which look like common clams and mussels, although they are not related).

In some areas of Devonian-age rocks, fish are actually the commonest fossils. In coal-mining areas the most common fossils are the Carboniferous-age plants that make up the coal. If you look at common non-shiny coal (not anthracite) in big lumps, you should be able to see twigs, stems, and fern-like leaves. In those days there were warm seas in many parts of the world; if you live near some Carboniferous limestones, you may find that they are full of big corals, brachiopods, sea lilies and other coral-reef life.

Permian and Triassic rocks were generally laid down on land, and fossils are not so easy to find. Marine rocks of Jurassic and Cretaceous age are often full of shellfish, such as oysters and the coiled ammonites which swam above the sea-bed. These rocks may also contain the bones of sea reptiles. Dinosaur fossils are mainly Jurassic and Cretaceous in age, but they are not so easy to find as the shellfish that lived at the same time. This is because dinosaurs lived on land and they are preserved as fossils only if their remains ended up in the sea or at the bottom of a lake.

More recent rocks, of Cainozoic age, often contain very rich shell-beds if they were laid down in the sea or in lakes, and fossil fish and mammals may also be found in some places.

Unusual fossils

The rarest and most exciting fossils show us the soft parts of long-dead animals. These include insects in amber. Amber is solid resin which oozes from the bark of certain kinds of trees, and it sometimes traps insects and other small creatures. Every detail can be seen, since the bodies of these animals have not decayed at all.

Very rare are whole mammoths which have been deep-frozen for thousands of years in the frozen soil of Siberia. The flesh is often preserved very well, and scientists have even eaten mammoth steaks! We can tell from preserved mammoth hair that they were coloured red.

Trace fossils are fossilized animal droppings, footprints and burrows. Many museums have fossil trackways made by dinosaurs. We can tell how big the track-maker was, and how fast it was going. Commoner trace fossils include burrows and crawling marks made by ancient sea creatures. These can be very important to scientists, since they are often the only evidence they have that certain soft-bodied animals existed.

◄ Mammoth fossil found in Siberia, Russia.

▼ Long-legged fly trapped in amber, from Germany.

Looking for fossils

Anyone can look for fossils. You can find out from your library or your teacher, what kinds of local rocks there are in your area and how old they are. Your local museum should have specimens on display. The best place to find fossils is on a beach, where the waves keep washing away the rock and bringing out new fossils. Other places to look include cuttings at the sides of roads or old quarries. However, never go to any of these places alone, since they can be very dangerous. Make sure you go with a parent or a teacher.

◄ Fossil crinoids, a kind of sea animal which still exists today and is known as a sea lily. They were attached to the sea-bed by a long stalk.

◄ Fossil of an ammonite found in the rocks at Lyme Regis, Dorset, England. Ammonites were alive in the Jurassic period of geological time.

Coal

The heat energy which is produced by burning a piece of coal comes originally from the Sun. This is because coal is the fossilized remains of plants. These plants grew with the aid of the Sun's energy millions of years ago.

How coal is formed

It takes a very long time for trees, ferns and other plants to turn into coal. Dead plants usually decay completely, leaving no remains. But in swampy areas the process of decay is very slow. Dead plants pile up and form the spongy material known as peat. In places such as Ireland and Scotland, peat is dug out of the ground and dried. It burns quite well but is rather smoky.

Many of the peat bogs that formed in geological time were close to the sea, often near the mouths of rivers. The sea washed sand, clay and gravel over the peat. The weight of these materials made the peat sink, squashing it. Other layers of peat and sand formed above the first. Eventually the sand, clay and gravel turned into rock. The peat was so compressed that it, too, became hard, turning into coal.

Types of coal

Not all coal is hard, black and shiny. Peat that is not too compressed produces lignite, a coal that is soft and brown. This is mined in large quantities in Germany, Russia and Australia. The most common kind of coal is called bituminous coal. This black coal is easy to use because it does not crumble like lignite, and it burns easily. Where the peat is highly compressed, it produces anthracite, which is hard and black. Anthracite is difficult to set alight, but once on fire, it burns slowly with very little smoke.

Opencast and drift mines

Where coal occurs near the surface, it is dug up by opencast quarrying. Giant machines cut a trench through the soil and surface layers of rock to reach the coal seam. Smaller mechanical shovels then dig out the coal. Although opencast quarrying is cheap, it can ruin the landscape.

Coal seams are often found one above another, sandwiched between other rock strata, rather like a layer cake. Where part of a coal seam reaches the surface of the ground, for example on the side of a hill, miners can tunnel horizontally straight into the seam. This type of mine is called a drift mine.

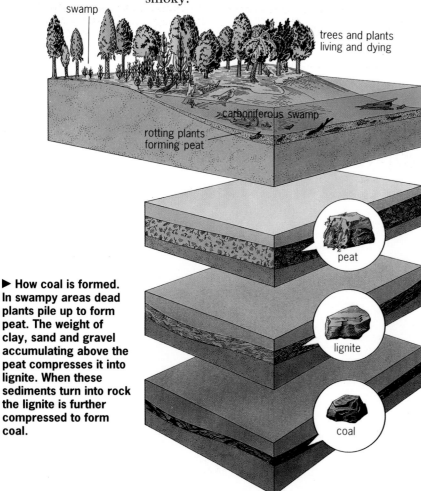

swamp

trees and plants living and dying

carboniferous swamp

rotting plants forming peat

peat

lignite

coal

▶ How coal is formed. In swampy areas dead plants pile up to form peat. The weight of clay, sand and gravel accumulating above the peat compresses it into lignite. When these sediments turn into rock the lignite is further compressed to form coal.

Shaft mines

Most coal seams are found far below the surface. To reach them, two vertical shafts are dug. One of the shafts is used to carry the miners up and down, and the other to lift the coal and to provide a second way out in an emergency. The shafts are also used to ventilate the mine. One pumps in fresh air and the other extracts stale air and dangerous gases. Galleries or 'roads' are cut from the shafts to the coal face.

Nowadays most coal is mined mechanically. A machine slices the coal away from the face of the seam with a rotating cutting-head. A conveyor belt carries the coal away to the foot of the shaft. The coal is then loaded into the lift and taken to the surface. The roof of the mine immediately behind the coal face is supported by hydraulic props made of steel. When the coal reaches the surface, it has to be washed and sorted.

Different sizes and qualities of coal are used for various purposes.

Uses of coal

Some coal is simply burned as fuel, not only in household fires but also in power-stations, to produce electricity. Much coal is also turned into coke. In this process the coal is baked instead of burned. The gases given off are collected and used to make a number of important chemicals.

When coke is made, coal tar and ammonia are also formed. Many chemicals are present in coal tar, and these are used to make a wide range of products, including plastics such as nylon, explosives, the wood preservative creosote, and even cosmetics and medicines such as aspirin. The ammonia is made into fertilizers. Coke is a valuable smokeless fuel. It is also used in making iron from iron ore.

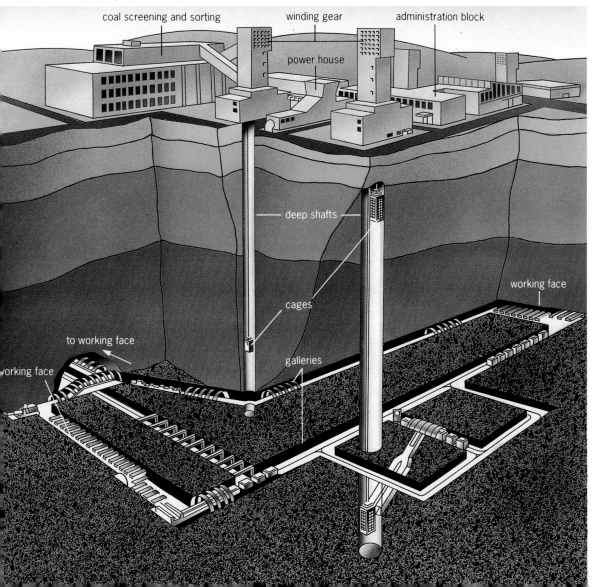

coal screening and sorting · winding gear · administration block · power house · deep shafts · cages · to working face · galleries · working face · working face

◀ **Main features of a deep-shaft mine. Today machines do most of the work. Electric lights have replaced oil-burning lamps, and electric railways and conveyor belts do the work once done by ponies.**

Oil

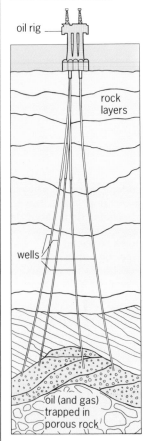

▼ Oil is often found deep in the ground under thousands of metres of rock. Sometimes the rocks are under the sea. At sea, a single oil rig can control the flow of oil from several wells.

oil rig

rock layers

wells

oil (and gas) trapped in porous rock

Oil is the remains of tiny plants and animals which lived in the sea millions of years ago. When they died, they sank to the bottom and were covered by layers of mud and sand. Over the years, the remains were buried deeper and deeper, and the mud and sand turned to rock. Heat, bacteria and the weight of rock above slowly changed the remains into oil (and natural gas). Some seeped upwards to the surface and escaped. But some was stopped by layers of solid rock. Here it stayed, trapped in the holes in porous rock, rather like water in a sponge. In some places, earth movements bent the rock layers, creating huge natural traps for oil. The geological name for this oil is petroleum. Today, oil companies search for it under the sea-bed or under land once covered by sea.

Searching for oil

The search for oil starts with geologists who decide where the most likely rock layers are to be found. Next, geophysicists survey the area. They measure slight changes of magnetism or gravity which might give clues about the rocks underneath. Then they set off explosions to send shock waves down through the rocks. By recording the echoes reflected back, they can work out how the rock layers are arranged. If the signs are good, test drilling are made to see if there really is oil there.

► This is a semi-submersible rig, which is held down when its legs are flooded.

Drilling for oil

At the drilling site, a framework tower called a derrick is erected. Under the derrick, a drill is passed through the centre of a circular metal table. A diesel engine turns the table and this turns the drill. As the hole deepens, the drill is lengthened by adding drill pipes from above. The pipes are raised and lowered by cables running from the top of the derrick. The cutting end of the drill is called the bit. It has hardened steel teeth, and there may also be diamonds on it for cutting very hard rock. To keep the bit cool, watery mud is pumped down the drill pipes and out through holes in the bit. As the drill goes deeper, the hole is lined with steel. If oil is found, it can flow up through the drill hole. The hole becomes an oil well.

Where oil is found under the sea-bed, the derrick and drilling equipment are built on an oil rig. Some rigs have living quarters for the workers, though for safety reasons they often live on a separate platform. Some rigs float, while others rest on legs on the sea-bed. Some are semi-submersible. They are floated to the site. Then their legs are flooded with water until they rest firmly on the sea-bed. The choice of rig depends on the sea-bed, the depth of water and the likely weather conditions. The rigs used for natural gas are similar to those used for oil.

Pumping and transporting oil

When a successful drilling has been completed, valves are fixed to the well head (top) to control the flow of oil. Then the derrick is taken away and a pump is fitted. Oil flowing from the well is called crude oil. It is carried by pipeline, tanker ship, rail or road to a refinery.

At the refinery, the crude oil is broken down into a variety of substances, which can be combined with other things to make many new materials, including plastics, chemicals and artificial fibres. Tarry bitumen for road-making comes from the residue left at the bottom of the refining tower.

REFINING OIL

Crude oil is a mixture of several substances. At the refinery, these are separated. Separation starts with a process called fractional distillation. This takes place in a tall tower. The crude oil is heated until most of its liquids boil and turn into vapours. The vapours rise up the tower, cooling more and more as they rise until they become liquids again.

The different substances turn into liquid at different temperatures, so they flow out of the tower at different levels. The separated substances are called fractions. They include petrol (gasoline), kerosene and the oils used as diesel fuel and lubricants. Gas, similar to natural gas, is also produced.

Some of the liquids from the distillation tower are too thick and heavy to be much use as they are. But they can be broken down by a chemical process called cracking to make more petrol and light oils.

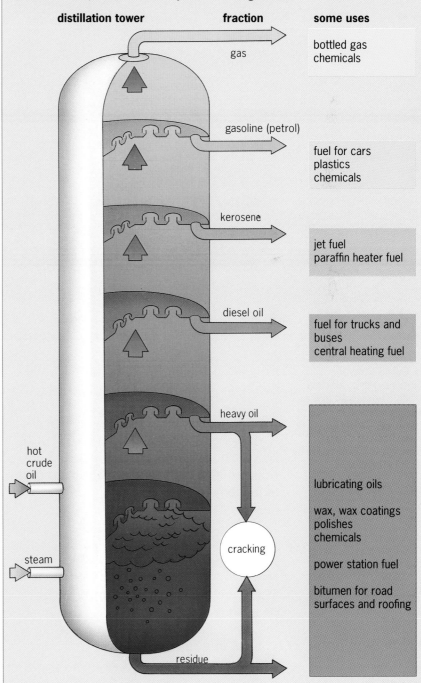

distillation tower   fraction   some uses

gas — bottled gas, chemicals

gasoline (petrol) — fuel for cars, plastics, chemicals

kerosene — jet fuel, paraffin heater fuel

diesel oil — fuel for trucks and buses, central heating fuel

heavy oil

hot crude oil

steam

cracking

residue

lubricating oils

wax, wax coatings, polishes, chemicals

power station fuel

bitumen for road surfaces and roofing

Geological time

▲ Grand Canyon National Park, Arizona, USA. The canyon is the deeply eroded course of the River Colorado. The river has cut down through 2 billion years of rock layers.

the enormous changes that had taken place through time in the fossil record.

It soon became clear to these early geologists, men such as James Hutton (1726–1797) and Charles Lyell (1797–1875), that layers of sedimentary rocks such as mudstones, sandstones and limestones could be arranged in sequences. The oldest rocks were usually at the bottom of the pile, and the younger ones at the top. They saw the layers of rock like the layers in a great cake. If you look at a high cliff made from sedimentary rocks, you can often see this pattern.

The first rule of relative dating, then, was to assume that the oldest rocks in an area were the lowest ones in the sequence. The second, very important rule was that the fossils in these rocks could give their age. It soon became clear that the history of the Earth could be divided into hundreds of time units, each represented by a different set of fossils. Particular fossils, or groups of fossils, could be used as guides to the age of rocks in any part of the world. These techniques of relative dating are still used in the oil industry to identify the ages of rocks from boreholes. Even a tiny sample with some small fossils can pin down the age to within a few million years.

How old is the Earth? How long ago did the dinosaurs live? Did human beings live with the dinosaurs? These are all questions to do with geological time: the way in which Earth scientists divide up the record of the rocks. The amounts of time involved are huge: millions of years, and even thousands of millions of years.

Relative dating

Geologists (Earth scientists) realized a long time ago that the Earth was very ancient. They could see great piled-up layers of rocks that had been laid down slowly on the ocean floor. They could see the effects of huge movements in the Earth's crust which lifted great masses of rock up, folded them, or turned them over. They could see

Absolute dating

Absolute, or precise, ages can be obtained from particular kinds of rocks which crystallized rapidly with some radioactive materials in them. Radioactive elements decay, or break down, at a constant rate. The rate of this decay is known. For example, it would take 4,510 million years for half of a sample of uranium-238 to break down to become lead-206. By measuring the exact proportions of these two elements in a rock sample, geologists can get a measurement of how long ago the crystals formed.

Era	Period	Epoch	Millions of years ago	Evolution and events
CAINOZOIC	QUATERNARY	Holocene / Pleistocene	2	The Great Ice Age. Modern humans appear
CAINOZOIC	TERTIARY	Pliocene / Miocene / Oligocene / Eocene / Palaeocene	66	Many mammals appear / Alpine earth movements create Alps, Himalayas and Rockies
MESOZOIC	CRETACEOUS		135	Dinosaurs die out / Chalk deposited
MESOZOIC	JURASSIC		205	Many dinosaurs
MESOZOIC	TRIASSIC		250	First dinosaurs and mammals
PALAEOZOIC	PERMIAN		290	Continents move together to form the giant landmass Pangaea
PALAEOZOIC	CARBONIFEROUS		355	Great coal swamp forests
PALAEOZOIC	DEVONIAN		412	Caledonian earth movements. Ferns and fish
PALAEOZOIC	SILURIAN		435	First land plants
PALAEOZOIC	ORDOVICIAN		510	Animals without backbones
PALAEOZOIC	CAMBRIAN		550	Trilobites. First shellfish
PROTEROZOIC	PRECAMBRIAN			First jellyfish and worms. Algae

The oldest rocks are the Acasta Gneisses in northwest Canada. They are 3962 million years old.

Some mineral crystals are older still. Zircon crystals from the Jack Hills of Perth, Western Australia, are 4276 million years old.

The earliest human-like remains were discovered near Lake Turkana in northern Kenya. They have been dated at 2·9 million years old.

Human type footprints have been found in volcanic ash 3·6 million years old.

The geological clock

If the Earth's 4600 million years is squeezed into the 12 hours of a clock face, we can see when different events took place.

2.52 the oldest rocks found today were formed.

4.20 the earliest life forms, simple bacteria and algae, appeared.

10.30 many-celled animals became common.

11.18 the first land plants appeared.

11.25 the age of dinosaurs began.

11.50 birds and mammals have replaced the dinosaurs. The first flowering plants appear.

11.59½ man-like animals appear.

Plate tectonics

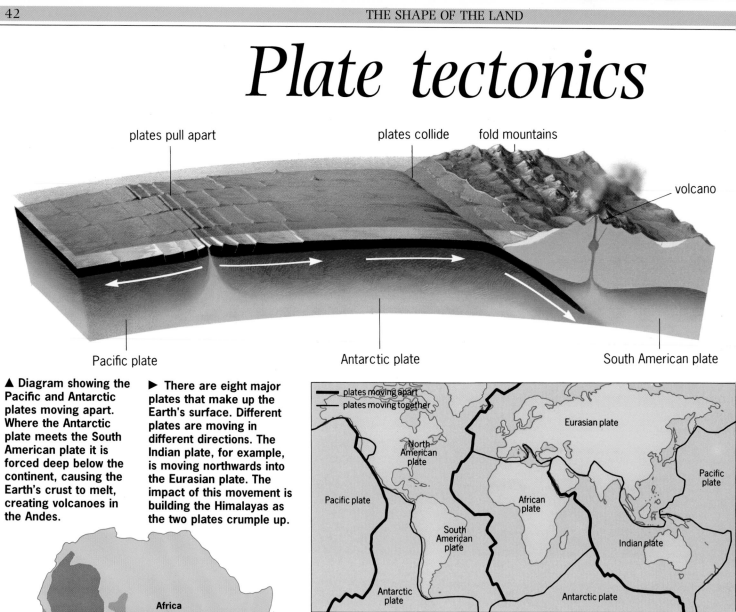

plates pull apart

plates collide   fold mountains

volcano

Pacific plate

Antarctic plate

South American plate

▲ **Diagram showing the Pacific and Antarctic plates moving apart. Where the Antarctic plate meets the South American plate it is forced deep below the continent, causing the Earth's crust to melt, creating volcanoes in the Andes.**

▶ **There are eight major plates that make up the Earth's surface. Different plates are moving in different directions. The Indian plate, for example, is moving northwards into the Eurasian plate. The impact of this movement is building the Himalayas as the two plates crumple up.**

plates moving apart
plates moving together

North American plate

Eurasian plate

Pacific plate

Pacific plate

African plate

South American plate

Indian plate

Antarctic plate

Antarctic plate

Africa

India

South America

Antarctica

Australia

Colours show similar types of rock

Lystrosaurus fossils

Cynognathus fossils

Glossopteris fern fossils

▲ **The ancient continent of Gondwana as it probably was before the continents moved to their present positions. The fossils and rock types that have been used as evidence of this original 'supercontinent' are shown.**

The explanation of what has happened to the surface of the Earth is known as the theory of plate tectonics. The skin of the Earth is divided into plates. These are great rafts of the Earth's crust and mantle rocks, about 70 km (45 miles) thick. They float on the slushy part of the mantle (the main inner layer) and move slowly over the Earth's surface. They move only a few centimetres each year, but over millions of years this has caused continents to split apart and collide.

How they move

On some parts of the ocean floor, plates pull apart. New crust is formed when molten material from undersea volcanoes rises to plug the gap. There are other plate boundaries where one plate edge is pushed under another. These areas also have volcanoes and earthquakes. In other places plates scrape sideways, splitting the crust.

The evidence

Identical fossils of ferns and reptiles have been found in South America, Africa, India and Antarctica. These species could not have crossed the oceans, so they must have lived on the same land mass at some time in the past.

If you cut out the continents on a world geological map and put them together like a jigsaw puzzle, you can match rock formations. It is unlikely that this is a coincidence. The rocks were formed when the land was a single continent. This continent, which geologists call Pangaea, split and the separate parts drifted to their present positions.

Continents

The large masses of land on the Earth's surface are called continents. Most geographers agree that there are seven, but there is sometimes disagreement about where one continent ends and another begins. Europe is the smallest continent. Some people insist that it is really part of Asia and that the two together should be called Eurasia. There is no clear division between the two. The Ural Mountains and the Ural River are usually taken to be the boundary. India is so large and has such a distinctive shape that it is called a sub-continent. Oceania is counted as a continent although made up of Australia, New Zealand and other Pacific islands.

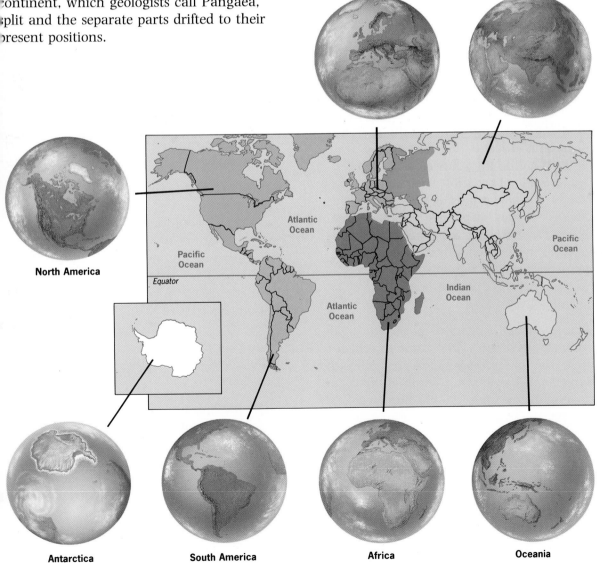

Europe **Asia**

North America

Antarctica **South America** **Africa** **Oceania**

Atlantic Ocean
Pacific Ocean
Pacific Ocean
Equator
Atlantic Ocean
Indian Ocean

◀ **This map shows the position of the continents. The circular pictures show how each continent appears from space.**

Grasslands

Grass covers more than one-fifth of the Earth's land surface. Many different types of grassland fill the vast areas lying between forests and deserts. They are called different names in different parts of the world. There are vast savannahs in East Africa, and veld in southern Africa. In North America there are the rolling prairies; in South America the dry pampas and chacos. In Central Asia are the desolate steppes, and in Britain the lush meadows and downlands. Grasses are hardy plants; they can live through disturbances such as fire, drought, and being cut by lawnmowers, and will grow again quickly.

Grassland wildlife

The world's grasslands are home to large herds of grazing animals. Australia has its kangaroos, while the North American prairies once supported vast herds of bison. Today, the main grazers are the prairie dogs, small rodents that feed on the prairie grasses, eating the roots as well as the seeds. Some of the most spectacular herds live on the savannahs of East Africa. They include giraffes, elephants and black rhinos. Herds of wildebeest, zebra and eland graze together here. Each animal eats different plants or parts of plants: some feed on grass, others on the scattered trees or shrubs.

Grasslands are dangerous places for grazing animals, as there is little cover in which to hide from predators such as lions, cheetahs, eagles and hawks. Only small animals can take refuge in underground burrows. Some animals have good hearing, others good eyesight or sense of smell. Grazing in mixed herds enables them to warn each other when danger is near.

The plentiful seeds and insects support large numbers of birds, such as the weaver birds of Africa and the kookaburras and brightly coloured budgerigars, parrots and cockatoos of Australia. There are also large grazing birds such as the Australian emus, African ostriches and South American rheas. Some birds also feed on the snakes and lizards that are common in grasslands.

Most grasslands have a host of tiny animals, easily overlooked. Earthworms, ants and beetles live among roots and leaves and in the soil. Termites build large 'skyscraper' homes of soil here.

People and grasslands

Humans have changed grasslands in many parts of the world. Three hundred years ago perhaps 60 million bison roamed the prairies of North America. Settlers from Europe shot bison for their meat and hides and by the beginning of this century fewer than 1,000 were left. During the 20th century the prairies have been changed even more dramatically. The grasslands have been ploughed up and wheat planted as far as the eye can see.

Grasslands have been ploughed to plant food crops in many other regions too, such as the steppes of Russia and Central Asia and the pampas of Argentina. In fact, wheat, barley and millet are all types of grass, too. Even so, the ploughing has not been completely successful. Grasslands do not always receive enough rainfall for crops. The drought in the North American prairies in the 1930s resulted in massive amounts of soil being blown away in the Dust-bowl.

Some parts of the steppes of Central Asia and Mongolia are still roamed by camels and horses. Some are wild, others are used by nomads, who move about with their animals and live in large round tents made of felt (called *yurts* or *gers*) which are warm in the cold winters. Much of South America's grasslands are used to graze cattle for beef.

▼ In Texas, USA, a prairie dog looks out for predators across the rolling grasslands. Prairie dogs live in burrows in large groups of a thousand or more.

▲ The North American prairies produce wheat and other cereals which are exported all over the world. Farming has been very successful and has changed the landscape. This photograph is of the Great Regina Plain, Saskatchewan, Canada. The silos are used to store grain before it is sold.

Cattle are raised on the pampas of Argentina; meat and hides are exported all over the world.

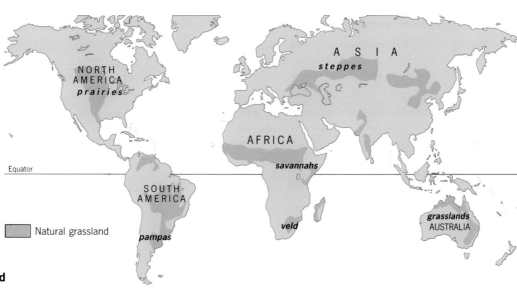

Natural grassland

▲ The steppes stretch across central Asia from Russia to Mongolia. The yurts (round tents) in this photograph are the homes of nomadic people who live on these vast plains.

▼ The savannahs of East Africa provide a habitat for herds of wild animals. On these dry grasslands you can see a herd of impala with zebra behind them.

Wetlands

Wetlands are damp, marshy areas where water lies on the surface in many places, often forming lakes or large pools. There are many different kinds of wetlands. Some are formed naturally, while others are the result of human activities. Wherever water is trapped by rocks or impermeable (water-resistant) soil, wetlands form.

Flood-plains and estuaries

Where rivers meander slowly over large flat flood-plains, the water often becomes trapped in pools and lakes, especially where the river spreads out to form a fan-shaped delta where it enters the sea. As the river meets the still waters of the sea, or a lake, it slows down and drops its load of sediment, forming muddy banks, especially on the inside of bends. River estuaries are usually muddy, and are often lined with saltmarshes, home to many invertebrates and wading birds.

▼ The Okavango delta is a vast area of marshland in the north-west of Botswana in southern Africa. This is natural wetland formed where the Okavango River divides into small streams.

Tundra wetlands

In the far north, over the vast treeless plains of the tundra in Canada, Alaska, Russia and Scandinavia, the soil is frozen not far below the surface. When the snow melts in spring, the water cannot drain away through the frozen soil, so it lies in great pools on the surface. Here mosquitoes and other insects breed, attracting millions of migrating birds from countries further south.

Mangrove swamps

In tropical countries the coasts are often fringed with mangrove swamps. Mangroves are trees adapted to live in salty, wet places. They have aerial roots which stick up from the mud like knobbly knees. These roots take in oxygen from the air in order to breathe, as there is very little oxygen in the wet mud in which they live. As the mangrove leaves fall into the mud they rot, enriching the mud with minerals.

Millions of invertebrates live in the rich mud of mangrove swamps, and fiddler crabs scuttle across its surface, scavenging the dead remains of plants and animals. Mudskippers also skitter over the mud on their front fins, breathing a mixture of air and water stored in their gill chambers. The warm, shallow waters form a nursery for the young of many ocean fish. Crocodiles or alligators lie in wait for them. The fish, frogs and other small animals attract birds such as storks, ibises and herons.

Ever-changing wetlands

The margins of wetlands are always changing. Where rivers enter the still waters of lakes or the sea, they drop their sediments, and the mud gradually builds up the banks. The original margin of the water is now dry land, and new species of plants start to grow there. As the land is built up further by mud, wind-blown soil becomes

apped in the new vegetation and other
species such as shrubs and trees can grow.
Along the coasts, saltmarshes gradually
extend seawards. Some wetlands are
constantly shrinking as they silt up. But
new ones form where rivers change course
or become blocked by landslides or dams.

Destruction of wetlands

In recent years, wetlands have been
disappearing at an alarming rate. Many
have been drained so that the land can be
built on or farmed. Others have been filled
in as rubbish dumps. Some have become
polluted as the rivers that feed them have
picked up pesticides and other chemicals
from farms and factories.

While the loss of wetlands has devastating
effects on the local wildlife, it can be
disastrous for humans, too. Wetlands often
act as buffers to river floods. They soak up
the water and let it drain out gradually, so
lessening the effect of the flood further
downstream. Where farmers have drained
lands for cultivation, often villages and
farms further downriver have been flooded
and crops lost.

Wetlands can also act like a water filter,
removing many of the impurities from the
water that passes through them. In some
places loss of wetlands has led to poor-
quality drinking water for local cities.

Wetlands protect the coast

At the coast, the loss of wetlands can be
much more serious. In many parts of the
tropics, cutting down mangroves for timber
or to make fish farms has resulted in severe
damage to villages further inland during
tropical storms and hurricanes. Expensive
artificial barriers and other forms of
protection have had to be built. The loss of
mangroves can also reduce fish catches, as
the fish lose their mangrove nurseries.

International importance

Destruction of wetlands in one part of the
world can have serious effects on the

wildlife somewhere else. The numbers of
small migrating birds reaching northern
Europe in spring have fallen since wetlands
just south of the Sahara have been drained
to provide grazing for cattle. The birds have
lost their last vital stopover point for
feeding and drinking before crossing the
Sahara, so many now perish on the
journey. Some important wetlands are now
protected by international agreements.

Managing wetlands

The conservation of wetlands presents
many problems. Simply leaving wetlands
alone is not enough. Many would gradually
silt up and disappear. Their water supply
must be carefully controlled. Sometimes
grazing animals such as sheep or horses
may be used to prevent tree seedlings
becoming established. If the areas
surrounding a wetland nature reserve
continue to be drained, the water in the
reserve will drain into these areas and be
lost anyway.

Tourists can sometimes bring in extra funds
for conservation. In the Everglades in
Florida, USA, many kilometres of wooden
walkways direct people along restricted
tracks which do not damage the habitat.

▲ The huge marshy
area of the Pantanal in
Brazil is criss-crossed
by river meanders. As a
river changes course,
old meanders form
large lakes rich in
wildlife.

Wetlands cover 8,500,000
sq km (3,280,000 sq
miles) – 6 per cent of the
Earth's land surface.

43 nations have signed the
international Ramsar
convention to protect the
world's wetlands since
1971. This has so far
saved only 200,00 sq km
(77,000 sq miles) of
wetlands.

Wetlands can remove 20 –
100 per cent of the heavy
metals from the water
passing through them.

Deserts

About ⅛ of the land area of the world is true desert.

The world's longest drought was broken in 1971 when rain fell in the Atacama Desert in Chile for the first time for 400 years.

Deserts are dry. Geographers say that a desert is an area which has less than 250 mm (10 in) of rain in an average year. But what is an average year? Rainfall is very unreliable in deserts and semi-deserts. A place may have heavy storms and floods one year, then no more rain for many years. It is impossible to draw the boundary of a desert. There is a very gradual change from almost no rain in the true desert to a short rainy season in the semi-desert.

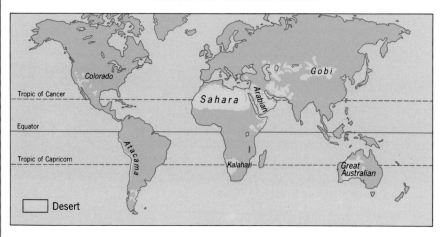

the clear skies allow heat to escape and it can become surprisingly cold. Dew sometimes forms by dawn. The deserts of Central Asia are not near the tropics. They are far inland, where the winds are dry. Deserts such as the Gobi are hot in summer but very cold in winter.

Sand, stones and rock

Only about a tenth of the Sahara Desert is sandy, and most other deserts have even less sand. Large areas are stony, such as the 'gibber' deserts of central Australia. Deserts contain mountains, some of which have been eroded into weird shapes, such as the flat-topped mesas of Arizona in south-west USA. Many of the strange-shaped rocks are the result of wind erosion. The wind can also pile up sand into dunes. As sand blows up the side of the dune and rolls down the steep face, the dune gradually moves, and may bury oases and buildings.

Storms and erosion

Water is important in shaping the desert. Even a little dampness on rocks helps break

▶ These dramatic rocks in Monument Valley, Arizona, USA, are the result of wind erosion. Sand blown by wind wears away the softer rocks, leaving the hard layers sticking out.

Hot and cold deserts

Most deserts lie near the tropics of Cancer and Capricorn. These are the hot deserts, where the Sun shines down relentlessly from a cloudless sky. But during the night,

▼ Sand dunes in the Sahara Desert. The name given to them in Arabic is *erg*, which means 'sand sea'.

...hem up when there are big changes in the temperature between day and night. Storms may not happen very often, but when they do, they sweep down the wadis (dry valleys) and flood over the land. Steep-sided gorges and great spreads of rock and gravel are the result of water erosion. When the storm is over, the water soon dries up, and some lakes become salt-flats.

People

Nomads who wander in semi-desert areas have developed special skills for finding water to survive. The Bushmen of the Kalahari live off the native plants and animals. The Bedouin of the Middle East graze herds of camels.

Today, there are probably more people than ever living in deserts. Once, the only perm- anent settlements were in the oases and in

◀ Desert vegetation in Arizona, USA. The tall 'saguaro' cacti in the background are often used by woodpeckers, who excavate holes for their nests.

the valleys of rivers like the Nile which flow through the desert. But people have made their own oases where valuable minerals are found. Deep wells, pipelines and even lorries supply water for settlements near the oil wells and mines in many desert areas.

Desert wildlife

Many desert animals survive by sleeping during the heat of the day and coming out to find food at night. Many live under- ground where heat and cold do not penetrate. Some 'aestivate': they sleep through the hottest part of the year. Plants, too, may become 'dormant' or inactive then, shedding their leaves to save water and surviving just as underground roots or stems, or as seeds.

Animals such as the kangaroo rat and gerbil can survive with little or no water. Desert toads and tortoises may store water in their bladders. Camels do not store water in their humps, but they can go for several weeks without drinking; they can drink enormous volumes of water in one go. Many desert plants store water in fleshy leaves. Cacti store water in their fleshy stems, which have thick skins to reduce evaporation. After rain, a cactus swells, then gets thinner as it uses the water. As its stem shrinks it forms deep pleats. As well as deep roots, it has many shallow ones to catch as much water as possible before it evaporates. Cactus leaves have evolved into spines. These reduce water loss and also protect the cactus from being eaten.

◀ In the Middle East, nomads known as Bedouin live in shelters that can easily be taken down, loaded onto a camel, and carried elsewhere. Deserts often do not provide enough food and water for permanent settlements.

The world's highest shade temperature was 58°C, recorded in Libya in 1922.

In the western Sahara, the difference in temperature between day and night can be more than 45°C.

The highest sand dunes that have been measured are in eastern Algeria. They are 430 m high.

Forests

In a rainforest there may be as many as 750 species of tree in an area of 1 sq km.

A forest is a large area of land covered mainly with trees and undergrowth. Some forests, such as the great Amazon rainforest, have existed for thousands of years. Although vast areas of forest have been destroyed by human activity, 20 per cent of the world's land remains covered in forest, both natural and specially planted.

Forests can grow wherever the temperature rises above 10°C (50°F) in summer and the annual rainfall exceeds 200 mm (8 in). Different climates and soils support different kinds of forests; for example, conifer forests grow in cold climates, and rainforests grow in the humid tropics.

► **Autumn in a deciduous forest.**

▼ **El Junque rainforest in Puerto Rico.**

The forest environment

Forests create their own special environments. The crowns of the tallest trees cast a shade on the forest floor. Where a large tree falls, the new patch of light encourages lush growth of vegetation, including the seedlings of more trees. There is little wind inside a forest. Trees draw up water from the soil, and some later evaporates from their leaves, so the air in a forest is still and moist.

The trees shield the forest interior from the full strength of the Sun, and also prevent heat being lost into the sky at night. So the temperature inside the forest does not vary so much as outside. The days are cooler, and the nights warmer, making the forest a sheltered place for wildlife.

The structure of a forest

A forest has a definite structure. The larger tree species have branching crowns which form an almost continuous canopy over the roof of the forest, allowing only small shafts of sunlight through. In tropical forests a few very tall trees, called emergents, grow through the canopy into the sunlight above. Below the canopy are smaller trees and the young saplings of taller trees. These form the understorey. Below them are shrubs and briars, and on the forest floor a layer of smaller shrubs and plants.

THREE DIFFERENT FOREST STRUCTURES

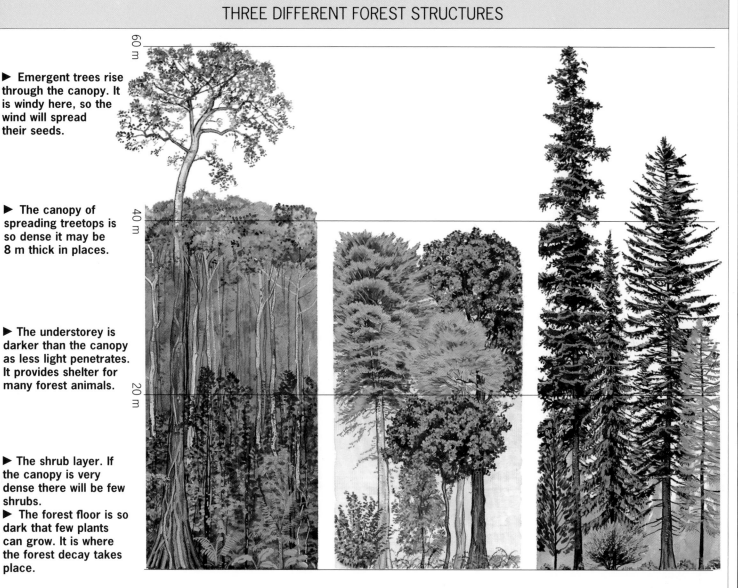

► Emergent trees rise through the canopy. It is windy here, so the wind will spread their seeds.

► The canopy of spreading treetops is so dense it may be 8 m thick in places.

► The understorey is darker than the canopy as less light penetrates. It provides shelter for many forest animals.

► The shrub layer. If the canopy is very dense there will be few shrubs.
► The forest floor is so dark that few plants can grow. It is where the forest decay takes place.

60 m
40 m
20 m

Tropical rainforest

Deciduous forest

Coniferous forest

The world's largest forest is coniferous. It is in Siberia and covers an area of 11 million square km.

The largest rainforest is Amazonia in South America. It covers an area of about 6·5 million square km.

In the warm humid climate of tropical forests, decay is so rapid that corpses can almost disappear in a few days.

Here temperatures are high and the air is moist all year round. The rainforest has such a dense tangle of vegetation that it is often difficult to distinguish the various layers. Plants, such as orchids, even grow on the trunks and branches of trees and fallen logs. There are so many different kinds of trees that there may be two or even three layers of understorey, formed by different species. Climbing figs, vines and lianas dangle from the branches high in the canopy. Herbs and shrubs carpet the forest floor. Leaves and fallen trees rot quickly in the moist, warm atmosphere, aided by many different fungi.

These forests are found in temperate climates. The main tree species are deciduous, which means that they shed their leaves in winter, or in the dry season. For part of the year the forest floor receives plenty of light, so herbs may flourish there. Many put out their leaves while the trees are still leafless. These forests have strong seasonal patterns of producing leaves, flowers, fruits and nuts. The understorey may include occasional evergreen trees and shrubs, such as holly and yew. Ivy, honeysuckle, clematis and other shrubs and herbs may climb or twine up the trees towards the light.

Coniferous forest is found further north and higher up mountain slopes than any other kind of forest. The main trees are conifers (cone-bearing trees) such as pines. Most are evergreen, with narrow leaves coated in shiny wax to reduce water loss, which would otherwise be high on windswept mountain slopes. They can also survive drought and the freezing of soil water in winter. Their branches slope downwards, so snow easily slides off. The dense canopy lets very little light through, so few plants grow beneath it, and there is a thick layer of dead leaves.

Forest life

Forests provide shelter and food for many animals. Leaves, flowers, fruits, seeds and nuts are food for insects, birds and small mammals, such as squirrels and mice, which in turn are food for larger birds and mammals. The moist forest soil has its own community: worms, centipedes, beetles, ants, and the eggs and larvae of many insects. Hollow trees, crevices under roots and bark, and the soil provide sites for nests and burrows. The forest is a noisy place, because visibility is poor among the trees and animals use calls and smells to communicate with each other.

Different layers of the forest have their own special communities. Birds, squirrels and, in tropical forests, monkeys roam the canopy. South American monkeys, mice, and some small marsupials and lizards use their tail as a fifth limb, curling it around branches.

Anteaters, lizards and bears have strong claws for climbing. Sloths and koalas have slow life-styles needing very little energy; they live by feeding on the leaves that have few nutrients, which other animals do not want. At night, bats hunt insects or forage for nectar and fruit.

Lower down are creatures that nest in holes, such as owls, toucans, starlings, woodpeckers and parrots. On the forest floor live burrow-dwellers such as badgers, stoats, mice, foxes and armadillos; foragers like wild pigs and deer; and large predators such as tigers, pumas and wolves. Snakes lie in wait among the dead leaves.

Rot and decay

Most of the food available in the forest is locked up in the bodies of the trees themselves. Dead trees leaves, twigs, flowers and fruits quickly rot. Woodpeckers drill holes in soft rotting trunks, and bark beetles excavate tunnels under the bark, allowing fungi to enter. Fungi dissolve the trees' tissues and absorb the nutrients. When the fungi die, they are broken down by bacteria.

The dead plant material gradually crumbles into the soil, where creatures such as worms, ants and termites, scavenging beetles, slugs and snails, soil fungi and bacteria set to work. As the once living material breaks down, the food it contained escapes into the soil, to be taken up by growing plants. Without rot and decay, forest soils would soon become too poor to support trees.

A similar process acts on dead animals. Vultures, crows, hyenas, jackals and other scavengers tear up the flesh, and flies lay their eggs in it. Their maggots clean the flesh off the bones.

Dense tropical rainforest in Queensland (above), and mangrove swamp in Kenya (left). Both provide habitats for numerous plant and animal species.

Moors and heaths

Moors and heaths are open treeless areas which were originally covered by forests. Over the centuries people cut down the forests to provide grazing for their animals. The grazing animals then prevented the trees from returning. Heaths are covered in low-growing heather-like shrubs, while moors are made up of coarse grasses and sedges. If grazing is very heavy, heaths turn into grassy moors. Heaths can become very dry in summer and fires are common, caused either by lightning or by humans. Without fires or grazing, trees and shrubs would soon return to heaths and moors.

Heaths

Heaths are found on acid soil. When the soil was covered in forest, the fallen leaves soon rotted into it, adding nutrients and making it less acid. Without the trees, the nutrients were soon washed away, leaving acid soil poor in nutrients. Heath plants grow very slowly, so do not need lots of nutrients. Very few bacteria can live in this soil, so plant material does not rot away, and forms peat.

Heaths are harsh, windswept places. Heathers and their relatives are low-growing woody shrubs. They have very small leaves covered in shiny wax to prevent them losing too much water by evaporation. Heathland plants produce juicy berries such as cranberries and blueberries, popular for making jams, jellies and pies.

Heaths are often used for grazing sheep. The tender young heather shoots make good fodder, and farmers often deliberately set fire to the heath from time to time to encourage new growth. Grouse rearing is also a profitable use for heaths.

Moors

Grassy moors are found mainly on slate and shale. Water cannot penetrate these rocks, so the soil here is waterlogged most of the

▲ **Windswept heathland near Tornahaish in Scotland.**

time. Waterlogged soils do not have many bacteria to rot down the plant material, so moors also build up peat. The peat itself tends to hold water, making the soil even wetter. The coarse grasses that grow here often form large tussocks (clumps), separated by wet boggy patches.

Wildlife

The shrubby heathers provide shelter and food for birds and small mammals. Insects thrive in moorland pools. Waders such as curlews nest on upland moors, feeding their young on the insects. Ground-living ptarmigan and grouse feed on insects and seeds. Small birds such as stonechats and warblers nest in the taller heather bushes; larger birds often nest on the ground. Heaths and moors are ideal hunting grounds for birds of prey such as kestrels, buzzards and kites. Lizards and snakes feed on the insects and the eggs of ground-nesting birds.

On heaths and moors there is little cover to hide in, so birds have to rely on camouflage rather than flight for protection.

In mountain moorlands, ptarmigan develop white plumage in winter for camouflage against the snow.

Cities

▶ You can see different parts of the city structure of Paris in this photograph: buildings of various sizes and heights, parks, roads and a bridge over the River Seine.

Metropolis and megalopolis
Sometimes the word 'metropolis' is used for a large city such as London. The London police force is called the Metropolitan Police. When cities grow into one another, as in the north-east of the USA, the word 'megalopolis' is used to describe the supercity.

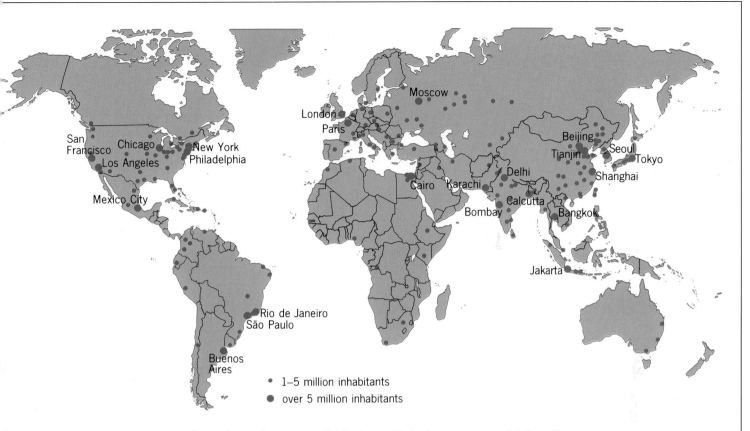

1–5 million inhabitants

over 5 million inhabitants

A city is an important town where lots of people live and work. It is likely to be bigger than other towns and may have some things small towns do not have. Cathedrals, for example, are usually found only in cities.

People and transport

About one third of the world's people live in cities. Some cities have very large populations, including a few that have more than 10 million inhabitants. Most of the people work in the city but it can take them a long time to travel from home to work. Roads become so congested that traffic is very slow moving. To get round this problem some cities, including London, Paris, Prague, New York, Tokyo and Moscow, have underground trains (subways).

Buildings

All cities are built up. This means that most of the available space has to be used to provide homes, business premises and civic buildings including city halls, schools and hospitals. Space is saved by building very

tall blocks called skyscrapers which will tower above older buildings. Many cities have a mixture of ancient and modern buildings because they have grown and developed over a very long time.

Structure

If you were to take a map of a city and colour in all the homes in one colour, all the shops another and so on for every different type of building, you would probably see a pattern. This pattern is called the structure of a city. For example, there are usually more houses and flats around the outside of a city than in the middle. The centre is often full of offices and shops, although there are also shops in among outer areas of housing. The main roads in and out of your city may show a pattern: leading to other cities, or avoiding high hills, for example. Factories may be together on one side, not in the centre.

Growth rates

Cities are always changing. Bus stops are moved, new shops may open, buildings are demolished and new ones built in their

▲ This map shows all the cities in the world with a population of over one million people. The cities which are named have more than five million inhabitants.

The largest city in the world is Mexico City, with a population of 12,932,116. This is the estimated population under the city administration. If you include the people in the suburbs as well, the population is about 18,748,000.

▲ The planned city of
Brasilia in Brazil.

Underground transport
The New York Subway has
the most stations (466);
London Underground has
the most track (more than
400 km); and the Moscow
Metro has the most
passengers (more than 6
million a day).

The oldest capital city in
the world is Dimashq
(Damascus) in Syria.
People have lived there for
the last 2500 years.

The city with the greatest
area is Mount Isa, in
Queensland, Australia. The
City Council administers an
area of 40,978 sq km.

place. The number of people living in a city
also changes. In many developing countries
the cities are growing very fast. But in
Europe and North America, some cities are
not growing, and may even be losing
people.

Capital cities

Every country has a capital city. This is
where the government of the country can
be found. If you want to know the capital
city of any country, just look up the
country in *Countries at a glance* and you
will find it. The capital city is often the
most important in the country, with the
most people and more businesses, shops
and factories than anywhere else. This is
not always so, because some countries
have built their capitals specially, such as
Canberra in Australia and Brasilia in Brazil.

Planned cities

Sometimes governments decide to build a
new city in a new place. Often these
planned cities are built especially to be the
capital. Brasilia, the capital of Brazil, was
founded in 1960. Many of the modern

buildings are for the government, but there
is a new cathedral, with a round roof
designed to look like Christ's crown of
thorns. In the Central American country of
Belize a new capital city called Belmopan
was built some way inland. The old capital
Belize City, was on the Caribbean coast,
and suffered badly from hurricanes.

Population figures

Large cities generally have two population
figures. A smaller figure shows the number
of people living in the area controlled by
the city's local government. The figures
used here are for the population under the
city administration, the metropolitan area.
Metropolitan area populations are often
used to decide which are the world's larges
cities. It is very difficult to compile
population figures that include the city plus
its suburbs and parts of other towns that
belong to the wider city area. The world's
three largest metropolitan area populations
are approximately:

Tokyo   19,040,000
New York City 18,000,000
São Paulo   16,832,285

THE WORLD'S LARGEST CITIES

Mexico City: Population 12,932,116

Mexico City is the capital of the country of Mexico. More people live in Mexico City than in any other city in the world. It is very large and very crowded. Air pollution is a serious problem in the city because of fumes from cars, buses and factory chimneys. The government has tried to tackle the problem, for example by banning people from using their cars one day a week. People can travel on electric trolley buses that don't produce exhaust fumes, or they can use the Metro. This is an ancient city: Aztec ruins have been discovered next to the cathedral, and north of the city are temples and pyramids over 1000 years old.

São Paulo: Population 8,490,763

São Paulo is the largest city in Brazil but is not the capital (which is Brasilia). The city centre is modern and wealthy, with skyscrapers, wide streets and spacious parks, but thousands of poor families live in the *corticos* (slum areas) without running water and proper sewage systems. The city has grown very rapidly this century because of immigration from Germany, Italy, Japan, and other parts of Brazil. The local people make textiles, clothing, footwear, motor vehicles, plastics and other goods. These goods as well as agricultural produce are exported through the Atlantic port of Santos.

Tokyo: Population 8,156,000

Tokyo, the capital of Japan, is on Honshu island, the largest of the islands which make up Japan. The Sumida River flows through the city and links with a network of canals. The transport system is very efficient and the city's flourishing industries use the most advanced technology. In 1923 an earthquake and fire destroyed half the city. Tokyo was again severely damaged by bombs in World War II. As a result most of the buildings are modern, but some fine Buddhist temples have survived.

Seoul: Population 9,645,000

Seoul is the capital of South Korea. Most buildings in the centre have been put up since World War II, particularly the enormous stadium built for the 1988 Olympic Games. Seoul dominates the commercial, educational, industrial and political life of South Korea, and contains one fifth of its population.

Moscow: Population 8,967,000

Moscow, the largest city in Europe, is the capital of Russia. The most famous part of the city is the Kremlin, originally a walled citadel. Here, 15th-century walls surround four cathedrals, several large palaces and the main government buildings. Beside the Kremlin is Red Square and the great 16th-century Cathedral of St Basil. Outside its historical centre Moscow is a modern city. There are wide avenues crossed by ring roads and lined with concrete blocks of apartments. The Moscow Metro, opened in 1935, has splendid, spacious stations.

New York: Population 7,262,000

New York is the largest city in the USA but is not the capital (which is Washington). It is built at the mouth of the Hudson River, and extends across Manhattan Island, Long Island and the mainland. The city is famous for the Statue of Liberty, its skyscrapers (especially the Empire State Building and the United Nations Building), Central Park, Wall Street (the financial centre), Broadway (theatres) and Fifth Avenue (expensive shops). Dutch colonists bought Manhattan for $24 of trinkets in 1626 and named it New Amsterdam. The English seized it from the Dutch in 1664 and renamed it New York. The city expanded as immigrants arrived in the late 19th and early 20th centuries.

London: Population 6,770,000

London is the capital and largest city of the United Kingdom. It lies on the lower reaches of the River Thames, which led to Roman invaders settling there some 2,000 years ago. The area of the Roman settlement is now an international financial centre, known as The City. London has a thriving tourist industry, attracting visitors to the shops, theatres and museums in the West End; and to historic buildings such as Buckingham Palace, Westminster Abbey, St Paul's Cathedral and the Tower of London.

▼ **Tokyo**

▼ **São Paulo**   ► **Moscow**

Farming

Most types of farming produce food for people to eat. Farmers make the best use they can of natural resources (such as soil and climate) to produce crops and rear animals. Different types of plants and animals need different conditions to grow well in, so there is a variety of types of farming around the world.

Dairy farming produces milk, butter and cheese from cows that graze in grassy fields. Dairy farms are usually quite close to large cities so that fresh milk can quickly reach people's kitchens.

Mixed farming involves both crops and livestock. The main area is the cornbelt of the midwest USA. Here farmers grow corn to feed to hogs (pigs) and cattle. Oats and hay are also grown as feed, as well as other crops such as soy beans and wheat. Mixed farming is found in Europe, too, in a region that stretches from northern Portugal and Spain across France, Germany and Poland and into Russia. In Britain mixed farms are found from Devon across the counties of the Midlands.

Mediterranean farming is found in areas with a Mediterranean climate where winters are mild, summers long and dry and rainfall is quite low. These areas are around the Mediterranean Sea, and also in California, Chile, South Africa and Australia. Winter crops include wheat, barley and broccoli. Summer crops include peaches, citrus fruits, tomatoes, grapes and olives.

Shifting cultivation is a common type of farming in many tropical countries. It is different from settled farming because shifting cultivators raise crops in a place for only as long as the soil allows the crops to grow well. After a year or so in one place the farmer moves on, chops away the natural vegetation from another area, and leaves the first plot to return to its natural state. Shifting cultivation is practised in the tropical forests of Central and South America, Africa and south-east Asia. Farmers grow maize, rice, manioc, yams, millet and other food crops.

Pastures and cattle ranges

Much of the beef in hamburgers eaten in North America comes from cattle that graze in Central and South America. To expand cattle-ranching, tropical forests have been cut down to provide grasslands for 'hamburger cattle'. Cattle also graze on natural grasslands such as the pampas of Argentina, where cattle have been herded by 'gauchos' on horseback for more than a hundred years.

In countries where intensive farming is practised, such as Britain, some cattle are not only fattened on pastures, they are also injected with drugs that make their bodies produce more meat. Where this is not done

► Dairy cows grazing on a farm in Somerset, south-west England. The mild climate and rich grass make south-west England just right for dairying, and the area is famous for its milk, cream, butter and cheese.

e farming is 'extensive' rather than
ntensive', as quite large areas of grassland
e needed to fatten one cow. In parts of
st Africa where grasslands are not good
ough to feed cattle all the year round,
rmers have to move their herds with the
asons to find new grazing.

Farm animals

eep, cattle, pigs, chickens and goats are
l farm animals. Sheep are kept both for
eir meat and for their wool. A farmer or
epherd leaves sheep to graze on grass-
nds. Dogs often help to round up the
eep and to protect them and their lambs
om wild animals such as wolves and
gles. Sheep are often reared on grasslands
at cannot be used for other types of
rming because they are too steep or too
y. Lamb and mutton are popular meats in
any regions of the Middle East. You
ould rarely see pigs kept on farms in the
liddle East, however, because most people
ere are Muslims and do not eat pork.
hickens are found on farms in many
egions of the world. In western Europe and
orth America large numbers of chickens

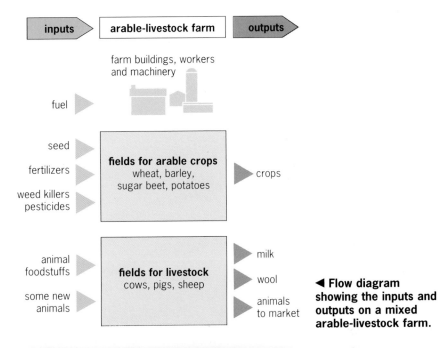

► Flow diagram
showing the inputs and
outputs on a mixed
arable-livestock farm.

◄ Battery hens on a
factory farm. Chickens
that are allowed to
roam free, unlike these,
are said to be 'free
range'. Some people
believe that factory
farming is cruel to the
animals. Others say
that it makes food
cheaper in the shops.

are kept indoors in row upon row of small
cages, often never seeing the light of day.
The farmer feeds these 'battery hens' each
day and collects their eggs.

Intensive farming

Battery hens are an example of intensive
farming: farmers organize their animals and
crops to get the maximum food from them.
Intensive farming uses a lot of machinery
to make it more efficient. Tractors are used
to plough fields and plant seed, and chemical fertilizers make plants grow stronger,
while pesticides kill pests and herbicides kill
weeds. In North America the wheat belt,
stretching from the USA into the Canadian
prairies, is a large area of intensive farming
where the wheat is harvested with combine
harvesters.

▲ Rows of tea bushes on a large plantation in Japan. The leaves are often picked by hand, although machines are increasingly used. The land here is used very intensively. Tea is an important and valuable cash crop.

Grain farming

Grain (cereals) is the most important food source for most people in the world. The main types of grain are wheat, corn (maize), and rice. Grain is eaten in all sorts of ways. Wheat is eaten mainly as pasta in Italy and as bread in North America and elsewhere. The USA, Argentina, Australia and Russia are the main world areas of wheat production. Farming in these countries is mostly intensive. Farmers use machines, fertilizers, pesticides and herbicides on large wheat fields. This intensive farming means that the amount of grain produced from a hectare of field (its 'yield') in North America is over four times that produced from a hectare in Africa. The USA is the world's biggest exporter of grain. Nearly every African country imports grain. Grain is also used for animal fodder.

The Green Revolution

In the 1960s special international efforts were made to breed new crop varieties that would produce better yields. This would produce more food from an average field to feed the fast-growing populations of tropical countries. Scientists were successful in breeding high-yield types of wheat and rice. This has become known as the 'Green Revolution'. In India, China, Mexico, the Philippines and other parts of south-east Asia the production of these foods has risen quickly. Some of the new rice types, for example, yield three times as much rice per hectare as traditional types. They also grow more quickly, allowing two or three crops a year from land that used to produce only one crop.

There are problems, however. The new varieties of plants need fertilizers and pesticides if they are to grow well and resist diseases. Not all farmers can afford these chemicals, which also cause pollution of soil and water.

Organic farming

Farmers who choose to farm organically do not use chemicals on their land. Fertilizers, pesticides and herbicides can cause problems. They kill plants and animals that the farmer does not want to kill, and very small traces of the chemicals may be left in crops which may make them dangerous to eat.

Organic farming uses compost and manure from farm animals to fertilize the land, and other plants, such as garlic, to control insects. Animals kept on organic farms are allowed to roam in the open air and are not locked up in cages for long periods. Organic farming has grown in the USA and Europe as farmers have realized some of the problems of intensive farming. Many people believe that food grown organically tastes better and is safer than food produced by intensive methods.

Farms of the future

Some of the problems of poor weather can be solved by breeding special new crops that are not so spoiled by frost or certain pests. This sort of breeding, using 'plant genetics', is going to become more and more common in farming in the future.

Forestry

Every year the world consumes 3 billion cubic metres (100 billion cubic feet) of wood. About half of this timber is used as firewood, the only source of energy the poor can afford. Forest trees also provide fruits and nuts, spices, oils for cooking and industry, syrups, resins, varnishes, dyes, rubber, latex, kapok, fibre, insecticides and drugs, including antibiotics. Many of these come from rainforest trees, and there are probably thousands of kinds of rainforest trees with uses that have not yet been discovered.

Types of trees

There are three main types of forest trees. Softwoods are conifers (cone-bearing trees) such as pine, spruce and fir. Most are evergreen, with needle-shaped leaves. They are used to make paper, furniture and house frames. Hardwoods are broad-leaved trees such as oak, ash, maple, eucalyptus, teak and mahogany. They are used for furniture, panelling, flooring, fencing and shipbuilding. Palms and bamboos are found in warm climates. Palms produce coconuts and dates, while bamboos are used for paper, furniture and building.

Managing forests

There are two main types of forests - natural forests, and plantations, where trees are planted in rows. The trees are usually clear-felled: a given area of forest is totally stripped of trees. Foresters need to make sure there will still be trees to cut in the future. Trees that are cut down must be replaced.

The forest may be left to grow back by itself. This is called 'natural regeneration'. When original, 'virgin' forest is cut down, trees and shrubs soon colonize the new space. This new forest growth is called secondary forest. It has fewer species and its plants and animals are much less varied

than those of virgin forest. The first colonists are plants with easily-dispersed seeds which can cope with windy, exposed sites. Fast-growing species shade slower-growing ones, which then die out, so the plants gradually change. After a long period of time, the forest may turn back into virgin forest, but only if it remains undisturbed and there is enough untouched forest nearby to provide new seeds. Or foresters may grow new tree seedlings in nurseries, then plant them in the forest.

The importance of forests

Forests are not just important for timber. A forest acts like a giant sponge, absorbing rainfall and preventing it from running away into rivers. It then releases the water slowly by evaporation from its leaves. This helps to stabilize the rainfall of areas downwind from the forest. Forest cover also prevents the soil being eroded (worn away) and silting up rivers and lakes. By holding back the water, the forest prevents disastrous flooding further downstream.

About 1 per cent of all the trees in the world are cut down every year.

In the late 1980s an area of about 200,000 sq km (80,000 sq miles) was burned down every year, releasing 7,600 million tonnes of carbon dioxide, one of the gases responsible for the 'greenhouse effect'.

▼ Logs waiting to be processed at a lumber mill in British Columbia, Canada. The logs are floated down the river from the forest to the mill.

Fishing

The average world catch of fish is 92 million tonnes a year. The biggest fishing nations are:

Japan	12 million tonnes
Russia	12 million tonnes
China	10 million tonnes
Peru	6 million tonnes
Chile	6 million tonnes
USA	5 million tonnes

Fish farming is not a new industry. In the Middle Ages fish ponds were made in many parts of Europe. Some of the largest are in southern Bohemia (now in the Czech Republic).

▶ **Unloading fish from the net into the hold of a Russian trawler.**

▼ **These are the three main types of net used to catch most of the world's fish. The drift and purse seine nets are used to catch fish near the suface, trawls catch fish near the ocean floor.**

The fishing industry employs many people to catch the fish, process and sell it, and make and repair the fishing equipment. Fish is an important source of protein for humans, but at least 40 per cent of the world catch is used to make fertilizer or fish meal, which is fed to cattle. This is a wasteful use of fish, because the processing destroys much of its protein content.

Inshore fishing

Almost every coastal village in the world has some fishermen to supply its daily needs. Inshore fishing boats do not sail very far from the coast. The fishermen may use rods and lines or small nets. The boats often sail together in fleets, leaving before sunrise and returning in the evening. The fishermen usually sell their catch the same day. In some parts of the world the fish are caught at night. Fish are attracted to lights shining from the boats.

Deep-sea fishing

Most of the world's fish catch comes from the deep oceans, too far from land for ships to catch and return the same day. Deep-sea fishing ships are bigger than those that fish

inshore. The Japanese are great fish-eaters and Japanese fishing fleets catch more fish than those of any other country. They travel all over the world, to the seas of the North and South Atlantic, the Pacific, the

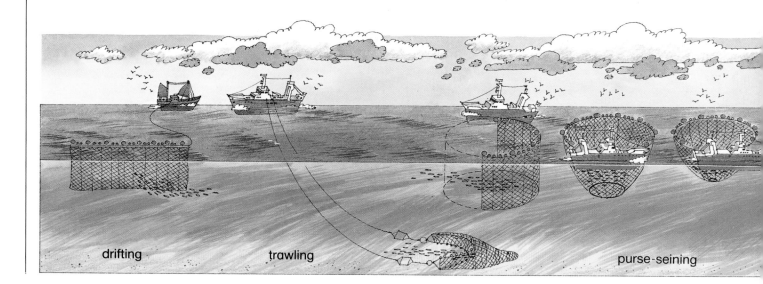

drifting   trawling   purse-seining

ntarctic and the Caribbean. The ships can stay at sea for months; many trawlers feed their catches into a 'factory ship' where the fish are processed while still at sea. In the factory ship the fish are cleaned, gutted and frozen.

The fish are caught in one of the three types of net shown below. Trawlers drag their nets along the sea floor to catch bottom-dwelling fish such as cod, hake and haddock. Purse seiners are used for fish such as blue-fin tuna, marlin and sailfish, which swim near the surface. Long-liners (lines of baited hooks up to 30 km long) are used for fish like yellow-fin tuna, which swim some way below the surface. The nets of some modern ships are too heavy to be dragged on board, so some have great pipes, like vacuum cleaners, that suck up the catch from the net into the hold while the net is still in the water.

Drift nets trail down from the surface, and may stretch for 50 km through the ocean, trapping anything that swims into them. Besides fish, thousands of sea turtles, seals, dolphins and even swimming seabirds drown in these nets every year. Old abandoned drift nets go on 'fishing' for decades. They have been nicknamed 'walls of death'.

Overfishing

In some oceans fishing fleets have caught too many fish. More and more ships have fished the same waters and the holes in their nets are too small to allow younger fish to escape and keep breeding to make the next generation. This 'overfishing' has meant that certain stocks of fish are now in danger, such as herrings in the north-east Atlantic and Pacific Ocean perch. Anchovies caught by Peruvian fleets used to be the largest single catch in the world, but since the early 1970s the numbers caught have declined dramatically. This is due partly to overfishing, and partly to changes in the ocean currents. Deep sea fish are found by echo-sounding: sending sound waves into the water and detecting the sound bouncing back from a shoal of fish.

Fish farming

China is the world's most important fish-farming country. Lakes, rivers and parts of the coast are stocked with fish which are bred just to catch and eat. This fish farming is also called 'aquaculture', and it is becoming more and more popular in many parts of the world. On many of Scotland's lochs (deep inland lakes) fish farms breed salmon. In coastal fish farms, sea-water fish, shrimps, crabs, lobsters, mussels and oysters can be reared, and are much easier to catch than at sea.

Artisanal fishing methods

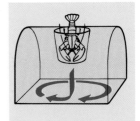

Lobsters and crayfish seek out dark places. They are caught in lobster pots baited with crabmeat.

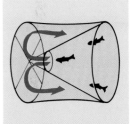

Fyke nets are used to catch fish swimming with the current.

Fijian rock-wall traps on gently sloping shores trap fish when the tide goes out.

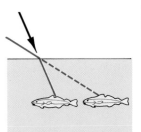

Light is bent at the water surface, so the spear fisherman has to aim at a point nearer than the fish appears to be.

Environmental problems

The 1980s were the warmest decade since reliable records began. If the Earth continues to warm at its present rate, sea-levels could rise by over 1 m by the year 2030. A 1m rise in sea-level would make 15 million people homeless in Bangladesh. It would flood ⅙ of Egypt's farmland. Large areas of London would be under water and Venice would be destroyed.

Burning fuels like petrol, oil, coal and natural gas puts about 6 billion tonnes of extra carbon dioxide into the atmosphere every year.

A typical family in the West, using electricity, central heating and a car, puts over 12 tonnes of carbon dioxide into the atmosphere every year.

There are 22 million road vehicles in Britain alone. Most of them are driven around carrying only one person.

▶ The Earth is warmed by the Sun. It loses heat by radiation, but some of this is reflected back by greenhouse gases in the atmosphere.

Before humans came on the scene, the world changed only slowly, except for local events such as earthquakes, volcanic eruptions and tropical storms. The climate warmed and cooled, new plants and animals evolved and became extinct in their turn, and sea levels rose and fell over periods of thousands, if not millions of years. But during the last two thousand years the rate of change has been dramatic. Forests have vanished, river courses have been altered, and large areas of natural vegetation have disappeared under farmland and cities. The delicate balancing processes of nature have been disturbed, and some of the results pose serious problems for the survival of the human race.

The greenhouse effect

It may be cold outside, but on a sunny day it can be hot in a greenhouse. Some of the gases in the Earth's atmosphere act like the glass in a greenhouse. Radiant heat from the Sun can pass through them to warm the Earth below. But the ground also loses heat by radiation. The 'greenhouse gases' reflect some of this heat back towards the Earth's surface and help to keep it warm. However, by burning fuels and forests, we are putting larger and larger amounts of these greenhouse gases into the atmosphere. As a result, the Earth is slowly warming up. This is called the greenhouse effect. Unless action is taken, it may cause great damage.

Greenhouse gases

Carbon dioxide is the main greenhouse gas in the atmosphere. Animals (like ourselves) give out carbon dioxide when they breathe,

while plants absorb carbon dioxide. In this way, animals and plants keep the atmosphere in balance and the amount of carbon dioxide stays the same.

However, our modern life-style is upsetting the balance. When we burn fuels in vehicles and power-stations, the exhaust gases put huge amounts of extra carbon dioxide into the atmosphere. In some countries vast areas of tropical rainforest are being burnt to clear land for development or cattle-rearing. This is causing a double problem. The burning is releasing more carbon dioxide and the Earth is left with fewer plants to absorb the gas.

There are several other greenhouse gases in the atmosphere. Methane is released by animal waste, swamps, rice paddy-fields, and oil and gas rigs. Nitrous oxide comes from car exhausts and from chemical

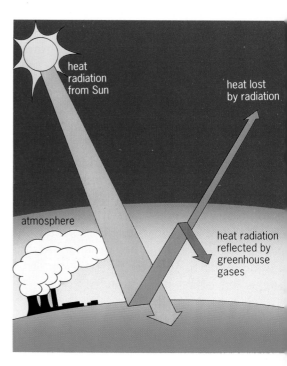

heat radiation from Sun

heat lost by radiation

atmosphere

heat radiation reflected by greenhouse gases

tilizers used to increase crop production. lorofluorocarbons (CFCs) have been used refrigerators, aerosols and foam ckaging. Amounts of CFCs in the mosphere are small, but these gases are 0,000 times more effective than carbon oxide at trapping heat.

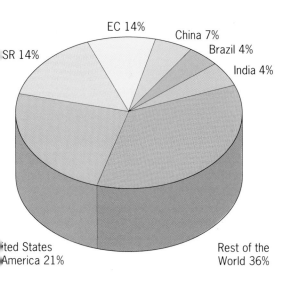

SR 14%   EC 14%   China 7%   Brazil 4%   India 4%

ted States America 21%   Rest of the World 36%

This shows how different regions of the world ntributed to global warming in the 1980s. The tions with advanced industry have caused most of e problem, and the United States is the main ender.

Solving the problem

1e problem with the greenhouse effect gan about 100 years ago when people arted using fuels like oil and petrol on a rge scale. On average, world temperatures ve risen by about half a degree (Celsius) er the last 100 years. They could rise by 10ther three degrees over the next 50 ars. This may not sound very much. But could cause droughts in some parts of the 0rld. If the polar ice-caps melt and sea- vels rise, many low-lying areas of the 0rld will be flooded.

ientists believe that the only way to slow e greenhouse effect is for us to produce ss of the gases which cause it. 0vernments are already trying to reduce e use of CFCs. We need to reduce our use fossil fuels like petrol, oil, natural gas and coal. We can develop heating systems and engines which burn fuel more efficiently. And we can build houses and offices which waste less heat, and transport systems which need fewer vehicles. We can also use sources of power that do not burn fuel and release carbon dioxide. Nuclear power is one possibility, but many people are worried about the dangers of using this. Other alternatives are wind, tidal, hydro- electric and geothermal power.

Air pollution and acid rain

Pollution is caused when what we do damages our surroundings. Factories, power- stations and motor vehicles pump waste gases, soot and grit into the air. The polluted air damages people's lungs. Some petrol has lead dissolved in it. The lead comes out in car exhaust fumes and it can cause brain damage in children.

The waste gases formed when coal, oil and petrol are burned include sulphur dioxide and nitrogen dioxide. These gases may dissolve in water in the atmosphere to form weak acids. They later fall to the Earth as acid rain, sometimes hundreds of kilometres from where they were formed. Much of the acid rain in Canada is caused by smoke from factories and power-stations in the USA; the acid rain in Scandinavia may come from Britain.

Acid rain attacks trees and other plants, and kills the fish and water animals living in lakes and rivers. Acid rain in soil changes certain metals, especially aluminium, into a poisonous form which damages tree roots. When the aluminium is washed into lakes it affects the gills of fish and kills them. Acid rain and polluted air also damage the bricks and stonework of buildings, and corrode the metalwork of steel bridges and railings.

▼ **Police in Milan, Italy, wear face masks as a protection against air pollution.**

▼ **Over the years, acid rain has slowly dissolved parts of this statue.**

▲ Algal growth forms an unpleasant scum on the surface of polluted Lake Constanta, on the Black Sea coast of Romania. When sewage and detergents run into rivers and ponds they provide a lot of excess nutrients (food), so tiny plant-like algae multiply in the surface water. These block out the light, so water plants die. When the algae die, they sink and rot, using up oxygen. Fish and other water animals die from lack of oxygen. This process is called eutrophication.

Preventing acid rain

Acid rain is difficult to control because it spreads so far. Building tall chimneys reduces the effect near the factory, but passes the pollution on to other areas. There are some types of coal and oil which contain very little sulphur. The waste gases of factories and power-stations can also be cleaned before they leave the factory, and cars can be fitted with devices to clean their exhaust smoke. These methods are all expensive.

The ozone hole

Ozone is a gas which forms a layer around the planet at about 20 to 50 km (6 to 30 miles) above the Earth's surface. The ozone layer prevents the Sun's dangerous ultra-violet radiation reaching the Earth where it would damage our skin and cause cancers. Many scientists are worried that the ozone layer is being destroyed by the chlorofluoro-carbons (CFCs) which are used in aerosols, refrigerators and freezers. These chemicals are also important greenhouse gases. At certain times of year the ozone layer becomes extremely thin near the north and south poles. Already skin cancers are increasing in Australia.

Many countries are trying to stop the pro-duction of CFCs and to find other chemicals to do the same thing.

Pollution of rivers and lakes

Air pollution affects rivers and lakes indirectly because it causes acid rain. But rivers may also be polluted directly. Some towns and villages pump untreated sewage into rivers, while factories sometimes release poisonous wastes into the water. Fertilizers and chemical pesticides used by farmers can also be washed by rainwater into rivers and streams. These pollutants can kill fish and other water animals and plants.

Pollution of the sea

Rivers eventually flow into the sea, carrying their pollution with them. But the sea can also be polluted directly. Some coastal towns and cities pump their sewage straight into the sea, while the oil that spills from oil tankers and oil rigs spreads over the ocean, killing seabirds, shellfish and other wildlife.

Heat pollution

All power-stations need huge amounts of water to cool them. This is taken from rivers, lakes or the sea. When it is returned the water is warmer than it was originally. Warm water does not hold as much oxygen as cold water, so it harms animals. It may kill fish and other water animals in the immediate area or prevent them from breeding.

Radioactive waste

Nuclear power-stations produce waste which is radioactive. Some of this waste is released by the power-stations into the air or water; some is stored. It can be carried long distances by the wind or by water currents. Many scientists worry about the long-term effects of this type of pollution on humans and wildlife.

Deforestation

More than 105,000 sq km (40,000 sq miles) of forest are being destroyed every year, most of them tropical rainforests. This would cover an area larger than Ireland and Wales together. Many rainforests grow on very poor soils, and there may well be more food locked up in the plants than there is in the soil. When rainforest is cut down and burned, the remaining soil is often too poor to support crops for long. With its cover of vegetation gone, the soil is easily washed away in the tropical rains, and it silts up lakes and reservoirs. With no trees to trap the rainfall, the area may even turn to desert. Areas downwind of the destroyed forest also become drier and do not support growth so well. Because more water runs freely off the soil, it causes flooding and mudslides further downstream.

Like all green plants, trees produce their own food by photosynthesis. They absorb carbon dioxide, an important greenhouse gas, and release oxygen. Forests are major suppliers of the world's oxygen. In the late 1980s an area of about 200,000 sq km (80,000 sq miles) was burned down every year, releasing 7,600 million tonnes of carbon dioxide.

Problems with roads

Roads help us carry goods and people around the country, but they can be unwelcome intrusions in wilderness areas. Roads built to bring logs out of forests are soon invaded by settlers who burn more forest to plant crops, not realising that the fragile forest soils will not stay fertile for long. Roads into wilderness areas make it easier for tourists to reach the area, putting pressure on wildlife and local resources. There is an increase in rubbish and problems of sewage disposal. More traffic arrives to supply the tourists with food, petrol and other goods, and soon it is not a wilderness area any more.

Dams

We all know that dams cause great devastation if they burst. But they cause other problems too. They upset the flow of water in the river downstream of the dam. Seasonal patterns of flow no longer exist. The water flowing out from the bottom of the dam is colder than the river water, which may kill many of the river inhabitants. Dams also trap mud and silt. Many river floodplains provide fertile soils for agriculture because the soils are renewed by silt brought down by the spring floods each year. Dams prevent fertile silt from reaching the floodplain.

Since the great Aswan Dam was built on the River Nile in Egypt, more land has been brought into cultivation by irrigation canals, but in the delta area people have had to start using fertilizer for the first time. The coastal fisheries have declined, because there are fewer nutrients in the water near the mouth of the Nile.

▼ Tropical rainforest being burned to make way for 'slash-and-burn agriculture' in the Amazon basin, Peru. The poor soils soon become infertile and are easily washed away in the tropical rains, so the farmers have to keep burning more and more forest. Wherever loggers and miners make roads into the forest, poor peasant farmers follow and begin burning the forest.

Atmosphere

The Earth is surrounded by a layer of air called the atmosphere. Imagine the Earth as an orange. Then the atmosphere is rather like the peel wrapped around the orange. The air itself is a mixture of gases, mainly nitrogen and oxygen. The atmosphere makes it possible for us to live on this planet.

Troposphere

This is the layer in which we live. It contains 90 per cent of the air in the atmosphere. Here clouds are formed and carried by winds over the Earth's surface. To help prepare weather forecasts, special balloons carrying instruments to measure weather conditions are sent up through the troposphere. The measurements are beamed back to Earth by radio. As you move up through this layer, the air becomes thinner and on high mountains there is not enough oxygen to breathe easily. The temperature drops and at about 10 km (6 miles) above sea-level it is always as cold as winter at the South Pole about −55°C (−67°F).

Stratosphere

Here the air is much thinner than in the troposphere below. Long-distance aircraft fly in the lower part of the stratosphere so as to take advantage of the lack of air resistance. Sometimes they are also helped by the high-speed 'jet-stream' winds of up to 300 km per hour (190 mph). Among the gases in the stratosphere is a type of oxygen called ozone. It absorbs much of the harmful ultraviolet radiation from the Sun.

Ionosphere

Within the ionosphere there are layers of particles called ions which carry electricity These layers are very important in bouncing radio signals around our planet.

Atmospheric pressure
The weight of the atmosphere is really quite considerable. There is more than a kilogram of air in each cubic metre of the air surrounding us. And we have the weight of all the air above pushing down on us. This atmospheric pressure is like having a kilogram pressing on every square centimetre of our bodies.

The total weight of the Earth's atmosphere is about 500 million tonnes.

The temperature of the troposphere gradually drops until it reaches the stratosphere, which is rarely more than 0°C. But the temperature of the ionosphere gradually rises as you move away from the Earth.

The atmosphere filters out harmful radiation from the Sun, and also helps to keep the Earth warm by preventing the Sun's heat escaping back into space once it has warmed the Earth's surface.

AURORA

People who live in northern countries, such as Norway, Canada and Scotland, sometimes see shimmering curtains of light and huge glowing patches in the night sky. The popular name for this display is the 'Northern Lights', but scientists call it the aurora.

The lights can be white, or coloured red, green, yellow and blue. The patterns they make are sometimes like rays from a searchlight, twisting flames or curtains blowing in a wind. Occasionally, an aurora can be seen from further south, in England, for instance. There are many old records dating back thousands of years that talk about these lights in the sky. Captain James Cook reported seeing an

aurora in the far southern sky when he was on one of his great voyages of discovery in 1773. We now know that they happen just as often in the southern hemisphere. But what causes them?

In the 18th century, people began to notice that there was a connection with the Sun. When there are big sunspots on the Sun, we are more likely to see an aurora. Atomic particles that burst off the Sun travel across space to the Earth. When they collide with atoms in our atmosphere, the different coloured lights are given off. This all happens very high up in the atmosphere, between 80 and 600 km (50 to 370 miles) above the ground.

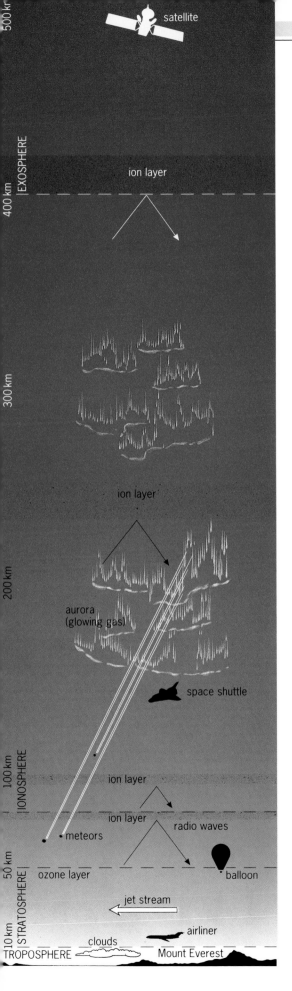

satellite

EXOSPHERE

500 km

400 km

ion layer

300 km

ion layer

200 km

aurora
(glowing gas)

space shuttle

IONOSPHERE

100 km

ion layer

ion layer

radio waves

meteors

50 km

ozone layer

balloon

jet stream

STRATOSPHERE

10 km

airliner

clouds

TROPOSPHERE

Mount Everest

A transmitter at a radio station sends radio waves up through the atmosphere and they bounce off layers in the ionosphere and return to Earth many hundreds of kilometres from where they started.

Exosphere

The exosphere has hardly any gas in it at all. There are only a few molecules of hydrogen and helium floating about. The exosphere is where the Earth's atmosphere really becomes part of space.

◀ This diagram shows the structure of the atmosphere. The heights given for the different layers cannot be exact. One layer merges into another and the altitudes vary depending on the time of year, the latitude, and activities of the Sun such as sunspots and solar flares. Space shuttles can orbit as high as 300 km.

Fog, frost and dew

All air has some water in it. Most of this water is a gas called 'water vapour'. Steam is really water vapour. If water vapour in the air is cooled, it becomes droplets of water and forms fog. During the night the ground cools down and lowers the temperature of the air above it. By early morning it may become cold enough for droplets of water vapour to form and make fog.

Air just above the ground has a lot of moisture in it. This is water from the soil and from plants themselves. When the air cools down at night, water vapour condenses on the cold grass to form dew. When the temperature of the air falls below the freezing point of water (0°C, 32°F), moisture in the air freezes into tiny ice crystals where the air touches blades of grass and branches of trees. This is frost.

▲ Water vapour in the air freezes when the air temperature falls below the freezing point of water (0°C, 32°F), forming frost on cold surfaces.

When fog forms in big cities it mixes with pollution in the air to form 'smog': half fog, half smoke. Londoners used to call such fogs 'pea soupers' because it was like trying to see through a thick pea soup.

Very cold air can damage our noses, fingers and ears if they are not well wrapped up. In severe cases of frost-bite the blood does not travel to that part of the body and it may die.

Climate

The climate of any place or region can be thought of as the yearly average of the daily weather. In Britain, for example, the weather often changes from day to day. This changeability of the weather is typical of Britain's climate. Other climates are much less changeable. A hot desert climate, for example, often has many days and weeks on end when the weather is hot, cloudless and sunny.

Elements of climate

When we talk about climate we use the words: temperature, rainfall, wind, pressure, humidity, cloud, sunshine, fog and many others. These are the elements that make up climate.

Tropical climates are the rainiest. The driest and hottest places in the world are in deserts. Timbuktu, in Mali, has an average annual temperature of 29°C (84°F). The coldest climates are at the poles. The Siberian town of Verkhoyansk, inside the Arctic Circle, has an average annual temperature of −17°C (1½°F).

Polar climates can be very harsh. In Antarctica winds of 145 km/h (90 mph) can blow continuously for 24 hours. When these winds pick up snow a 'whiteout' occurs, and it is possible to get lost even a few metres from camp. The poles are also very dry. Scientists think that it has not rained in the dry valleys of Antarctica for 2 million years.

Seasons

Another important part of climate is the way in which weather changes through the course of a year. Around the poles there are marked seasons, with a warm summer and a cold winter. Towards the Equator the seasons are less well marked. They are not cold and hot, but are often divided into a rainy season and a dry season.

The lowest recorded temperature was -89·2°C (-128·6°F) at Vostok, Antarctica.

In the northern Sahara Desert a place called Al Aziziyah in Libya has recorded a temperature of 58°C (136°F).

The rainiest town on Earth is Cherrapunji, in India, with an average rainfall of 11,437 mm (450 in) a year. That is over 17 times the average rainfall of London (UK).

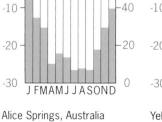

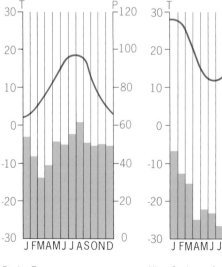

T:temperature (°C)   P:precipitation (mm)

Paris, France
Height 53 m

Alice Springs, Australia
Height 584 mm

Yellowknife, Canada
Height 208 m

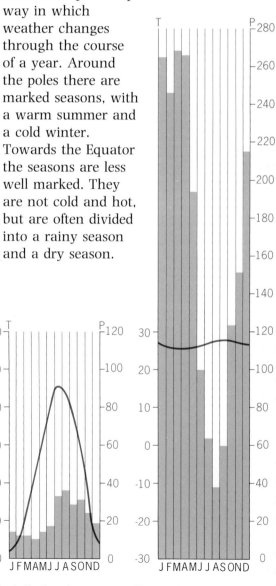

Manaus, Brazil
Height 44 m

Cyclones and anticyclones

yclones and anticyclones affect the
weather. When air warms up, it becomes
lighter and rises. When it cools, it becomes
heavier and sinks. A cyclone is an area of
low air pressure, which means the air is
rising. The air in a cyclone slowly revolves
around its centre. In an anticyclone the air
also revolves, but air pressure is high
because the air is sinking. Both cyclones
and anticyclones often move, but cyclones
usually move faster than anticyclones.

Cyclones

Cyclones have different names in different
parts of the world. In Britain they are
usually called 'depressions'. In tropical and
subtropical areas a cyclone with very strong
winds is called a 'hurricane' or 'typhoon'.
The weather in most cyclones is cloudy,
rainy and windy. A tropical cyclone seen
from a satellite is a great swirling mass of
clouds.

Cyclones form over the sea, and the violent
winds can do great damage as they roar
across the oceans and islands. When a
cyclone moves over large land areas it
gradually fades out. Hurricanes are given
names by forecasters who track their
progress and give warnings to people in
their paths. Hurricane-force winds range
from 120 km/h (74 mph) to over 320 km/h
(200 mph). But it is difficult to know exactly
what speed the winds reach because the
wind speed recorders are often blown away.

Anticyclones

A very large anticyclone stays over Siberia,
Russia, every winter which is called the
Siberian anticyclone. If you look at a
weather map of northern Asia, you will see
the Siberian anticyclone hardly changes its
position from October to March. The
weather in Siberia in winter is always
bitterly cold. The temperature often reaches
−30°C (−22°F). Other parts of the world also

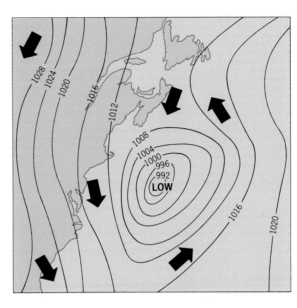

◄ In a cyclone the air pressure changes causing strong winds and stormy weather.

◄ In an anticyclone high air pressure remains stable over a wide area.

— isobars (lines joining places with equal air pressure) Numbers show air pressure in millibars.

➤ arrows show wind direction

have anticyclones in some regions in
certain seasons. The weather map over the
Sahara Desert often shows an anticyclone
in the winter months.

Anticyclones sometimes stay over western
Europe for a week. When this happens in
the summer the weather is hot and sunny.
There are few clouds in the sky and no
rain, although there may be fog in the
mornings. But in some parts of the anticyclone there may be summer thunderstorms when the air is hot and humid.

The highest air pressure
ever recorded was 1083·8
mb at Agata, Siberia, on
31 December 1968.

The lowest air pressure
ever recorded was
870 mb in Typhoon Tip
near the island of Guam in
the Pacific Ocean on 12
October 1979.

Depressions

Depressions are areas in the world's atmosphere where the air pressure at the Earth's surface is low. This means that the air is rising. A depression is a type of cyclone. In Britain the weather is constantly affected by depressions. These depressions form over the Atlantic Ocean and move eastwards across Britain and western Europe.

Cold and warm air

In the northern hemisphere, cold air from the Arctic meets warm air from the tropics over the Atlantic Ocean. These two masses of air meet in a series of swirls because the Earth is spinning. Each swirl is a depression. Cold air moves around each side of a parcel of warm air, gradually squeezing it upwards from behind. Depressions move towards the east. The line where two different masses of air meet is called a 'front'. A depression has two fronts. The first is the forward edge of warm air pushing against cold air. This is the 'warm

front'. Following it is the edge of cold air pushing against warm air. This is the 'cold front'.

Depressions over Europe

The warm and cold fronts are where most of the clouds form in a depression. A depression may take a couple of days to pass over the British Isles. Since it is made up of two blocks of cold air, like a sandwich around a warm sector, the temperature on the ground changes as the depression passes over. As the warm front passes over the temperature rises by one or two degrees and there is often low stratus cloud covering most of the sky, and a little rain. Behind the warm sector comes the second mass of cold air. This arrives at the cold front.

The cold front is pushing the warm air upwards. As the warm air rises, large clouds are formed, and there is heavy rain. Behind the cold front there are often big

Warm air meets cold air
The boundary between a mass of warm air and a mass of cold air is called a **front**. This is what a front would look like if the Earth stood still. But because the Earth is spinning, the air is always mixing where warm and cold air meet. This happens in a series of swirls, called depressions.

Warm and cold fronts
A wedge of warm air is trapped in the cold air. The warm air gradually rises over the cooler air because it is lighter. The warm front is marked on weather maps by a line with half moons on it. The cold front is shown by a line with triangles. The air rises fastest at these points and the result is usually bad weather.

Bad weather
You can tell when a depression is coming by the clouds. First comes cirrus, then alto-cumulus, followed by stratus. Last of all, at the cold front, come cumulonimbus, the big anvil-shaped rain clouds. Eventually the cold front catches up and the warm wedge of air is squeezed upwards.

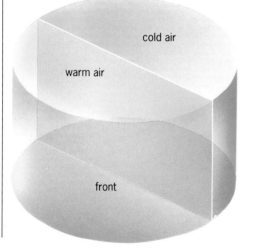

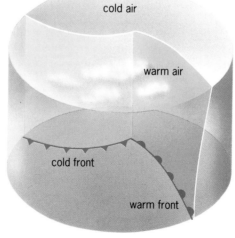

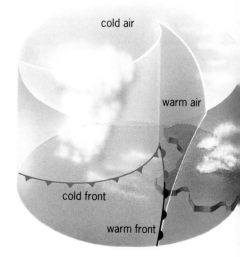

louds that look like blobs of cotton wool.
As the depression continues, the cold front
usually moves faster than the warm front.
Thus the cold front catches up the warm
front and pushes the warm sector up above
the ground. The two different masses of air
start to mix and the front is called an
occluded front.

Mediterranean depressions

Depressions also move through the region
of the Mediterranean Sea. Most of these
depressions form at the western end of the
Mediterranean. But they also move
eastwards. Some travel into the Middle East
and occasionally even as far as northern
India. Most of these depressions occur in
the spring or winter and bring rainfall.

Strong winds blow at the fronts of
depressions. When depressions pass over

◄ This dust storm in
South Australia has
been blown by the
strong winds at the
front of a depression. A
few minutes after this
photograph was taken
the photographer could
not see his hand in
front of his face
because of the dust.

deserts in any part of the world, these
strong winds can blow sand and dust from
the earth. A wall of dust is raised and
moves with the depression. It is hot and
dirty as the air is filled with dust.

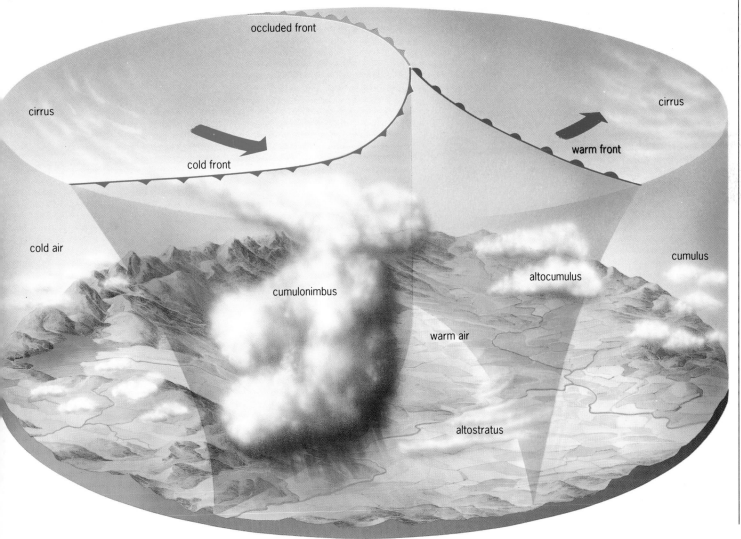

Clouds and rain

There is a lot of water vapour in the Earth's atmosphere. But since the temperature in the air is not always the same everywhere, the water vapour sometimes changes back to a liquid by condensation. Clouds are formed when water vapour condenses to become small droplets in the air. These droplets are so small they are not heavy enough to fall to earth as rain. They stay in the air and come together to form clouds.

Cirrus clouds look like wisps of hair, and usually lie 12 to 15 km (7–9 miles) up. At this height the temperature is always below freezing-point, so these clouds consist always of ice.

Cirrocumulus looks rather like the ripples in the sand on the seashore. It is often seen at the approach of fair weather after a depression. One form of cirrocumulus is the well-known 'mackerel sky'. The appearance of this cloud means that rain is on the way.

Altocumulus are layers of blob-shaped clouds arranged in groups, in lines or waves. In summer this cloud can often be seen in late evening or early morning.

Stratocumulus is a lower and heavier form of altocumulus.

Altostratus is a veil of even, grey cloud through which the Sun can be seen dimly. It gives a 'watery sky' and is an almost certain sign of rain, because it is usually caused by a current of warm, moist air flowing up over a 'cold front'.

► Different types of clouds form at different heights above the ground.
Low means below 3,000 m (9,800 ft).
Medium means from there to about 6,000 m (19,700 ft).
Cirrus clouds are the highest and may form up to 14 km (8 miles) above the ground.

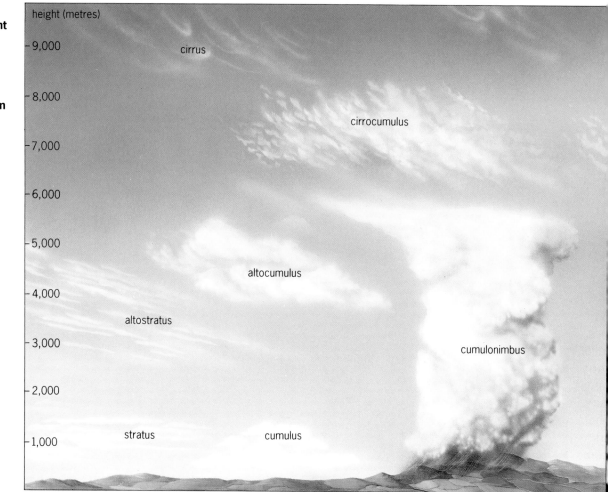

ratus is the lower type of cloud, often otting out all high ground. Many ountaineers and hikers have got into fficulties through the sudden appearance ' this cloud. It may thicken and turn to g, drizzle or rain.

imbostratus, the cloud which gives us ost of our heavy rain, also hangs low. It dark grey and threatening.

umulus are heavy, cauliflower-shaped ouds with flat bases. They are formed by onvection currents of rising air, warmed y reflection of heat from the Earth's irface.

umulonimbus is the thundercloud of ot, still summer weather.

Rain

ain is made up of thousands of drops of ater falling from clouds in the sky. Each rop contains many water molecules.

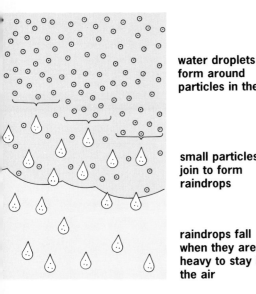

water droplets form around particles in the air

small particles join to form raindrops

raindrops fall when they are too heavy to stay in the air

Vhen air rises it becomes colder. Raindrops rm when the water vapour in the air is old enough to change from a gas to a quid, but this change is helped by very mall solid particles in the air. These specks ay be salt from the sea, dust or smoke. he water molecules collect around these nd soon a drop is formed. Raindrops fall om the sky when they are too heavy to tay in the air. They get bigger as they fall ecause other small drops join with them.

THUNDERSTORMS

Thunderstorms come from the biggest clouds in the sky. The clouds are called cumulonimbus and have a wide flat top and a narrower bottom. They bring heavy rain and often there is thunder and lightning.

Thunderstorm clouds have electrical charges. At the top they are positive, while at the bottom the charge is negative. The ground below a thunderstorm is also positive. When all these charges build up there is a lightning flash which very briefly lights up the sky. Thunder is the noise we hear when the air in front of a stroke of lightning is rapidly expanded because of the great heat.

Thunderstorms have bigger raindrops than other clouds. This is because the raindrops move up and down inside the cloud again and again, and each time they do another circuit inside the cloud they collect more water. When the drops are very big they are too heavy to stay in the cloud. Sometimes the air in the cloud is cold enough for the raindrops to freeze, and then they fall out as hailstones.

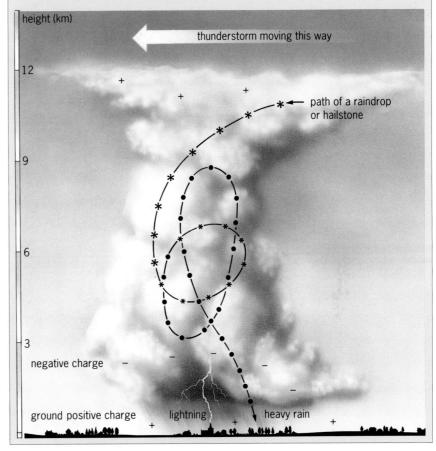

Rain is caused by different conditions. Mountain barriers may cause air to rise; the air above the heated ground may rise; or cyclones may form.

You may think of rain as something annoying that makes you wet. Farmers and gardeners like rain because without it their plants will not grow. But if there is too much rain it may cause a flood which can be damaging and dangerous.

Monsoon

The monsoon is the season of very heavy rain which occurs in most of south and south-east Asia. The word 'monsoon' means season. It was used by Arab sailors to describe the seasonal winds in the Arabian Sea. For half the year they blow from the north-east, and for the other half of the year from the south-west.

There are three seasons in the year in India and the neighbouring countries. From April to June, the weather is hot and dry. It gets hotter and hotter, until everyone longs for cooler weather and rain. Then, finally, the monsoon rains come. Long, heavy storms begin in June or early July. They water the earth and cool the air. It rains really hard for much of the next two months, and then there are two more months of hot weather and heavy showers. Then comes winter, the cool, dry season from October to March.

Farmers need the monsoon

The 'burst' of the monsoon can be forecast from year to year, almost to the exact day.

▼ In the Indian city of Calcutta, people scurry for cover from the torrential monsoon rains which mark an end to the hot dry weather.

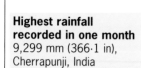

warm wet winds (June-July)

cool dry winds (October-March)

Annual rainfall
- very wet more than 2000 mm
- fairly wet 1000–2000 mm
- dry 500–1000 mm
- very dry less than 500 mm
- When the monsoon 'bursts' in a normal year

0 kilometres 600 (372 miles)   Indian Ocean

The rains are brought by warm, wet winds sweeping in from the southern Indian Ocean. They blow towards north-west India, where the temperature is hottest and the air pressure is lowest.

Throughout India and nearby countries, farmers wait anxiously for the day when the monsoon rains should arrive. They sow seeds in small fields and water them carefully, ready to plant them out in the wet ground. As soon as the baked earth has soaked up some rain, it is ploughed and prepared for the crops. If the monsoon rains come too early, or are exceptionally heavy, they may flood the land and wash away the young crops. But if the monsoon arrives late, or if the rains fail after the first few storms, then the young plants will wither away in the drought.

The countries of south-east Asia, eastern China and Japan also have monsoon winds and rains, and so does a small area of northern Australia.

Water cycle

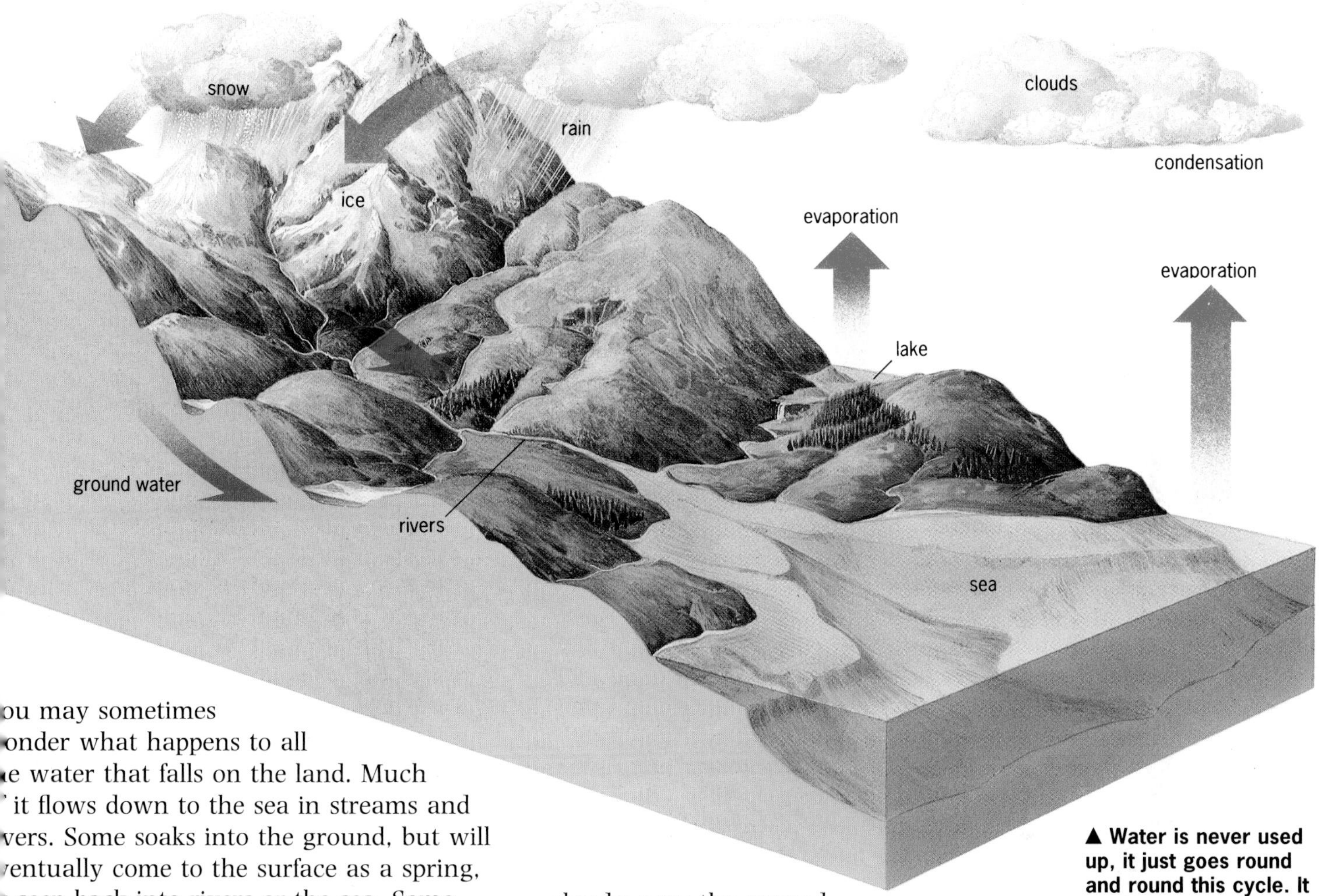

snow

rain

clouds

condensation

ice

evaporation

evaporation

lake

ground water

rivers

sea

ou may sometimes
onder what happens to all
e water that falls on the land. Much
' it flows down to the sea in streams and
vers. Some soaks into the ground, but will
rentually come to the surface as a spring,
· seep back into rivers or the sea. Some
ays as ice for a long time, but eventually
elts and also flows to the sea. But where
the rain and snow come from in the first
ace? The warmth of the Sun evaporates
ater from the sea, from rivers and lakes,
id also from the soil and plants on the
nd. The water is turned into water
apour, which is one of the invisible gases
at make up the air we breathe.

ir is moving about all the time. If it rises
gh up into the atmosphere, or if it comes
to contact with a cold area, it will cool
wn. Cool air cannot hold as much water
apour as warm air, so some of the water
apour will turn into tiny droplets of water.
/hen this happens on the cold window of
room at home, we call it condensation. In
e sky, the tiny water droplets form

clouds; near the ground
we call them fog, mist or dew. If the tiny
water droplets combine, they will fall to
Earth as rain, or as hail or snow if they are
frozen. The land is watered again.

The water cycle is kept going by the Sun's
energy which is used to evaporate water. In
some parts of the cycle the water is a liquid
(rain), in other parts it is a gas (water
vapour) or a solid (ice).

Pollution

Pollution is also carried in the water cycle.
Polluting gases that enter the air dissolve in
raindrops, making them acidic. This makes
the water of streams and rivers acid, too.
And this affects rocks and soils, plants, fish
and other animals.

▲ Water is never used
up, it just goes round
and round this cycle. It
is a remarkable fact
that the oceans hold
about 97 per cent of
the world's water. A
further 2 per cent is
frozen in the polar ice-
caps. This means that
only about 1 per cent of
the world's water is
going round the water
cycle at any
one time.

Wind

Wind is moving air. If you pick up a magazine and move it back and forth in front of you then you will feel the air on your face: a slight wind. Flapping the magazine needs energy. If you flapped for an hour your arms would get very tired. Wind also needs energy and this energy comes from the Sun. Sunshine warms and cools different parts of the air. When the Sun warms the ground, the air above it rises up and cooler air flows in to take the place of the air which has risen. This flowing air is the wind. You can feel wind on the ground and high up in the air. Weather forecasters use the scale of winds shown below, called the Beaufort scale.

Special winds

Many winds have names. Among the most constant winds are the 'Trade Winds'. These winds blow towards the tropics, where warm air is rising, and blow most of the year round. They were called the Trades in the times when much world trade was done by sailing ships. Ships always tried to avoid other areas in tropical ocean where winds hardly ever blow. These are called the 'Doldrums'.

A breeze is a light wind. Where the land and the sea meet, for example, you get land and sea breezes in summer. This is because the air is hotter over the land than the sea

THE BEAUFORT SCALE

◀ The Beaufort scale of wind speed, including calm (with no wind) and twelve different wind speeds.

Beaufort number	km/h	mph
0	below 1	below 1
1	1–5	1–3
2	6–11	4–7
3	12–19	8–12
4	20–28	13–18
5	29–38	19–24
6	39–49	25–31
7	50–61	32–38
8	62–74	39–46
9	75–88	47–54
10	89–102	55–63
11	103–117	64–73
12–17	over 117	over 74

0 Calm

1 Light air

2 Light breeze

3 Gentle breeze

4 Moderate breeze

5 Fresh breeze

6 Strong breeze

7 Moderate gale

8 Gale

9 Strong gale

10 Storm

11 Violent storm

12 Hurricane

y day and cooler over land than sea at ight. So the winds blow from the sea to nd during the day and from land to sea at ight.

n southern France a wind called the Mistral' often blows in March and April. It a cold, northerly wind, which can blow or several days. In the USA a wind called ne 'Chinook' is a warm, dry wind that lows down the eastern slopes of the Rocky Mountains. When the Chinook starts to low, the temperature rises quickly, ometimes by as much as 22°C (40°F) in ve minutes. These winds often make snow haw rapidly. The word Chinook means now eater' in a local Indian language.

Tornadoes

A tornado is sometimes called a whirlwind. t is a terrifying wind which destroys every-hing in its path. It looks like a violent, wisting funnel of cloud which stretches lown from a storm-cloud to the Earth.

A tornado is narrow and rushes across the and at speeds of 30 to 65 km/h (20 to 40 nph). The wind twists up within the funnel t speeds of up to 650 km/h (400 mph). It ucks up dust, sand, even people and nimals like a giant vacuum cleaner and lumps them where it dies out.

A tornado is quite different from a hurricane. It is smaller, faster and more iolent. Tornadoes are most frequent far nland, unlike hurricanes. A tornado that orms over the sea becomes a waterspout, sucking up water from the surface of the ocean. Most waterspouts are 5 to 10 m hick, and 50 to 100 m high. The highest on record was 1,528 m (5,014 ft).

Hurricanes

Hurricanes are violent storms with winds hat can blow at over 320 km/h (200 mph) ind bring torrential rain. They start over in area of warm sea, where they pick up a ot of moisture and hurtle towards the land. Such storms are called hurricanes in the Atlantic, but in the Pacific and Indian

▲ **Hurricane-force winds bend palm trees in the Caribbean.**

oceans they are often called cyclones or typhoons.

A hurricane may measure 400 km (250 miles) across. It is a swirling mass of winds which spiral upwards around the 'eye' at the centre of the storm. The eye of the hurricane may be 40 km (25 miles) across, and is a fairly calm area. But the strongest winds occur immediately around the eye. On the Beaufort scale, which measures wind speeds, the highest number is 12, used to describe a hurricane-force wind.

A place that experiences a hurricane has winds that increase in strength for several hours as the centre approaches. Pressure drops very low, and there is a lull before the terrible winds suddenly start again. As they die down, people come out to assess the damage. The winds can uproot trees, destroy buildings, and even lift up boats and cars and throw them around. Anyone caught outside does not stand much chance. Along with the heavy rains and destructive winds, hurricanes bring very high tides. Today, hurricanes can be tracked on satellite photos. Each one is given a name and warnings are issued in good time.

Tornadoes are especially common in the southern states of the USA, where they are known as 'twisters'.

The Coriolis effect
At the Equator warm air rises, creating a low pressure area. Cold air moves continuously from the North and South Poles towards the Equator. It does not flow directly north and south: the rotation of the Earth bends the winds to the right in the northern hemisphere, and to the left in the southern hemisphere.

Seasons

As the seasons change through the year, we notice the different things that happen in the world around us. In spring, the days get longer again after the winter and the Sun climbs higher in the sky each day. Because there is more sunshine and warmer weather, plants start to grow and it is a good time for sowing seeds. It is also the season when many animals have their young. Summer is the warmest time of the year. The higher the Sun is, the stronger the warming effect of its rays. Then in autumn the days shorten again, trees drop their leaves and the weather gets cooler as winter draws nearer.

▼ In June, the North Pole is tilted towards the Sun. It is summer in the northern hemisphere. In December, the South Pole is tilted towards the Sun. It is summer in the southern hemisphere.

► How the Sun's path through the sky, as seen from somewhere in the northern hemisphere, changes between the seasons. In summer the Sun gets much higher in the sky at midday than it does in winter.

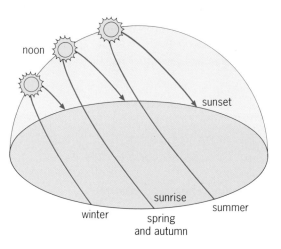

The planet Mars has seasons too. We can see its ice-caps grow in winter and shrink in summer.

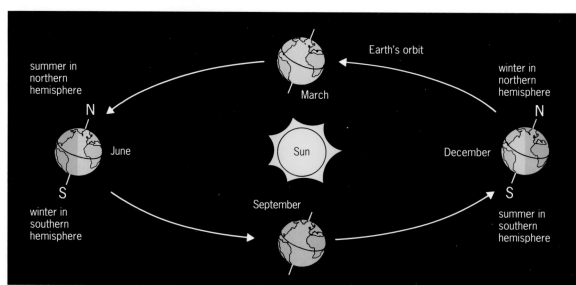

the night sky, the constellations you can e also change day by day, so the summer ars are quite different from the ones seen winter.

e seasonal changes are more extreme the rther you are from the Equator. Near the les, there are enormous differences tween the length of winter and summer ys, but it never gets really warm because e Sun is not very high in the sky, even in idsummer. Near the Equator, the number hours of daylight does not change much rough the year and the Sun is always gh in the sky.

Why we have the seasons

ur planet Earth is spinning around its is, an imaginary line going through the orth and South Poles. That is why we get y and night. At the same time, the Earth avels round the Sun once each year.

here are seasons because the Earth's axis tilted to its path round the Sun at an gle of 23½°. This means that the orthern hemisphere is tipped towards the un for half the year. During this time, om about 21 March to 21 September, aces in the northern hemisphere have ring followed by summer. At the same me, the South Pole is facing away from e Sun and the southern hemisphere is aving autumn and winter.

om September through to March, things e the other way round. The North Pole lts away from the Sun and the northern emisphere has autumn and winter while e southern hemisphere has spring and immer. The seasons in these hemispheres e always opposite. In July, when it is immer in Europe, it is the middle of the ustralian winter.

Equinoxes

quinox means 'equal night' and it is a ame given to two special days in the year hen the hours of daylight and darkness verywhere in the world are equal. The two ays fall on or near 21 March and 23 eptember. At midday on the equinoxes,

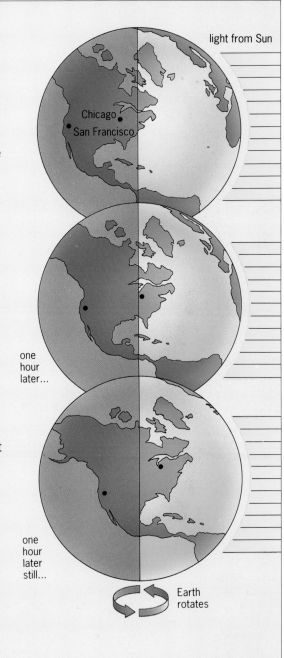

DAY AND NIGHT

Day begins when the Sun rises in the east and night comes when the Sun sets in the west. The Sun seems to move across the sky because we watch it from the Earth, which spins around once every 24 hours. The Sun is shining all the time, but we can see it only if our part of the world is facing towards the Sun.

As the Earth turns, places further and further west have dawn in turn. This is why there are time differences between countries. For example, the time in New York is five hours earlier than it is in London.

The length of day and night changes with the seasons. At places near the Arctic and Antarctic, the Sun never sets in the middle of summer and never rises in the middle of winter, so day and night can last for weeks.

light from Sun

Chicago
San Francisco

one hour later...

one hour later still...

Earth rotates

▶ Half of the Earth is in sunlight (daytime) and half is in shadow (night-time). As the Earth spins, Chicago moves from the shadow of night (top), through dawn (middle) into daytime sunlight (bottom). In San Francisco, sunrise comes later.

the Sun is directly overhead at places on the Equator.

Day and night are not the same length at any other times but change regularly with the seasons. The days when the number of hours of daylight is greatest and smallest also have a special name. They are called the solstices, and fall on about 21 June and 21 December. Sometimes these days are called midsummer or midwinter day.

March 21 is called the spring or vernal equinox. September 23 is called the autumnal equinox.

Solstice comes from the Latin meaning 'Sun standing still'. The solstices occur when the Sun reaches its most northerly and southerly position among the stars (namely latitudes 23½° north and south).

Weather

Rain, clouds, sunshine, wind: the conditions in the atmosphere and their day-to-day changes make up the weather. Knowing what the weather will be like saves money and can also save lives. Is it safe to climb the mountain tomorrow? Do we need to grit the roads tonight? Will it be dry enough to harvest the crop this week?

Weather recording

Weather stations on land need to be on open sites, away from buildings and trees which can influence the accuracy of readings. Observations must be precise and presented in a standard way so they can be compared with those taken at other sites. Most observations are made at ground level, but instruments can be carried into the atmosphere by balloons which transmit readings back to Earth by radio.

rain gauge

maximum and minimum thermometer

wind vane

and mercury. When the temperature rises the alcohol in the left-hand part of the thermometer expands and pushes the mercury up the right-hand part. A metal marker shows the position of the highest temperature. When the temperature falls, the alcohol contracts and the mercury is pulled in the opposite direction. Another marker shows the position of the mercury at the lowest temperature.

Wind direction is measured by a wind vane. The arrow of the wind vane always points to the direction from which the wind is blowing: the north wind blows from the north (not towards the north).

Wind speed is measured by an anemometer. The faster the wind, the faster the cups spin. A meter records wind speed in metres per minute.

Air pressure can be measured with a barograph. In the centre of this instrument is a metal box containing very little air. When the air pressure changes, the top of the box bends and this movement is recorded by a pen on a rotating drum.

Rainfall is collected in a rain gauge. Rain falling into the funnel trickles into a jar. At the end of the day the jar is emptied into a measuring cylinder marked in millimetres.

Maximum and minimum temperature are measured with a Six's thermometer. This is a U-shaped tube filled with alcohol

anemometer

barograph

sunshine recorder

Sunshine: the number of hours a place receives sunshine can be measured by a strip of sensitive card. The Sun's rays are focused onto the card by a glass ball and burn the card. The length of the burn shows the number of hours of sunshine.

Weather forecasting

Accurate weather forecasting depends on an accurate supply of information about weather conditions all over the Earth. Forecasters have to know what is happening over the whole planet, because today's weather in America could affect the weather in Europe a week later. Thousands of separate pieces of weather information are collected several times a day from weather stations on land, from ships and from aeroplanes. This data is fed into a

giant communications network called the Global Telecommunications System. The information is sent round the world at great speed by satellite, radio and cable. Weather forecasters in every country take from the system the data they need to make their own local forecast.

The information is first mapped on a chart which shows the overall weather situation. Using their knowledge of how the atmosphere behaves and with the help of computers which show what happened last time when there were similar conditions, meteorologists make their forecasts. These must then be turned into maps and descriptions of what the weather will be like for the next 24 hours, for newspapers, television and radio.

How accurate are the forecasts? They are becoming more accurate as time goes by, especially for short periods ahead. Weather is complex, and long-range forecasting is still difficult.

Weather and people

People have always made up sayings about the weather. Some are nonsense, but others are based on good observations. Many sayings are based on what the sky looks like. 'Red sky at night, shepherds' delight' is not always true, but 'Red sky in the morning, shepherds' warning' is usually reliable.

Scientists who study the weather are called meteorologists. Meteorology comes from two Greek words meaning 'study of what is high in the air'. (The word meteor originally meant anything unusual that appeared in the sky.)

▼ **The Global Telecommunications System supplies weather forecasters with accurate information on weather conditions all over the world.**

—— trunk routes for weather information

Washington

Moscow

Melbourne

Regions of the World

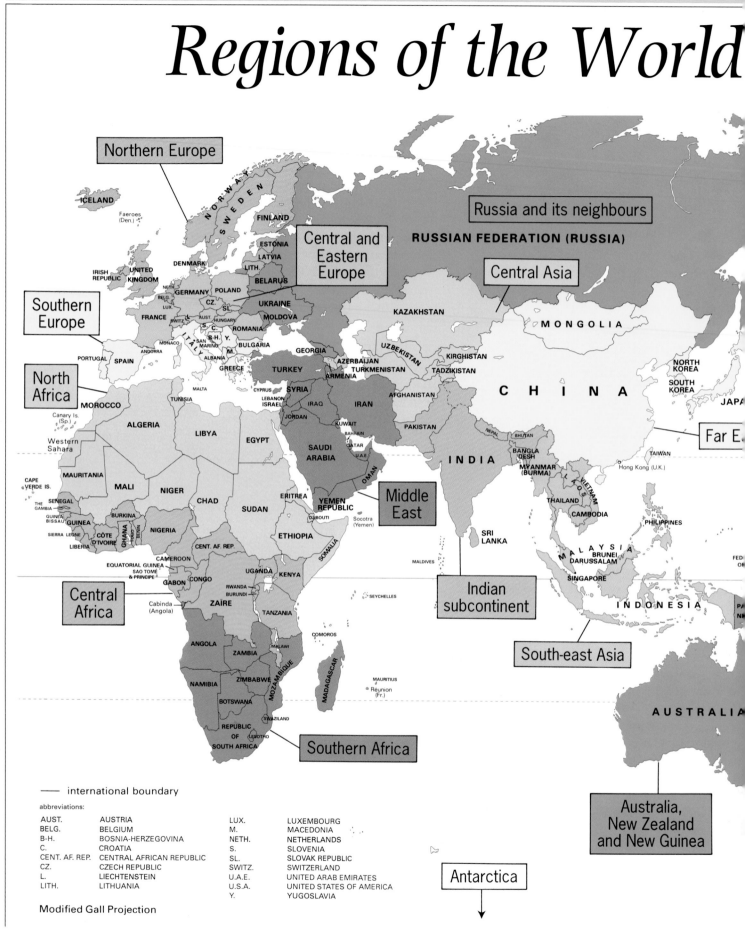

Northern Europe

ICELAND

Faeroes (Den.)

Russia and its neighbours

RUSSIAN FEDERATION (RUSSIA)

Central and Eastern Europe

Central Asia

NORWAY   SWEDEN   FINLAND

ESTONIA
LATVIA
DENMARK
LITH.
IRISH REPUBLIC   UNITED KINGDOM
BELARUS
NETH.   GERMANY   POLAND
BELG.
LUX.
CZ.   SL.
UKRAINE
FRANCE   AUST.   HUNGARY
SWITZ.   S.   C.
MOLDOVA
MONACO   SAN MARINO   ITALY   B-H.   Y.   ROMANIA
ANDORRA   M.   ALBANIA   BULGARIA
PORTUGAL   SPAIN

KAZAKHSTAN

MONGOLIA

Southern Europe

North Africa

MOROCCO

Canary Is. (Sp.)

Western Sahara

MALTA
TUNISIA
CYPRUS
LEBANON
ISRAEL
JORDAN

GREECE   TURKEY
SYRIA
IRAQ
KUWAIT
BAHRAIN
QATAR
U.A.E.

GEORGIA
AZERBAIJAN
ARMENIA
TURKMENISTAN

UZBEKISTAN
TADZIKISTAN

KIRGHISTAN

CHINA

NORTH KOREA
SOUTH KOREA
JAPA

IRAN

AFGHANISTAN

PAKISTAN

NEPAL   BHUTAN

Far E.

TAIWAN

ALGERIA   LIBYA   EGYPT

SAUDI ARABIA

Middle East

INDIA

BANGLA DESH
MYANMAR (BURMA)

Hong Kong (U.K.)

CAPE VERDE IS.

MAURITANIA

MALI   NIGER

THE GAMBIA
SENEGAL
GUINEA-BISSAU   GUINEA
SIERRA LEONE
LIBERIA   CÔTE D'IVOIRE

BURKINA
GHANA   TOGO   BENIN
NIGERIA

CHAD   SUDAN

ERITREA

YEMEN REPUBLIC

DJIBOUTI

Socotra (Yemen)

SRI LANKA

MALDIVES

THAILAND

LAOS   VIETNAM
CAMBODIA

PHILIPPINES

MALAYSIA
BRUNEI DARUSSALAM
SINGAPORE

FED. OF

CENT. AF. REP.

CAMEROON

EQUATORIAL GUINEA
SAO TOME & PRINCIPE
GABON   CONGO

ETHIOPIA

UGANDA   KENYA
RWANDA
BURUNDI

SOMALIA

SEYCHELLES

Indian subcontinent

INDONESIA

PA
N

Central Africa

Cabinda (Angola)

ZAIRE

TANZANIA

COMOROS

MADAGASCAR

South-east Asia

ANGOLA   ZAMBIA

MALAWI

MOZAMBIQUE

MAURITIUS

Réunion (Fr.)

NAMIBIA   ZIMBABWE

BOTSWANA

SWAZILAND

REPUBLIC OF SOUTH AFRICA   LESOTHO

Southern Africa

AUSTRALIA

Australia, New Zealand and New Guinea

——   international boundary

abbreviations:

AUST.	AUSTRIA	LUX.	LUXEMBOURG
BELG.	BELGIUM	M.	MACEDONIA
B-H.	BOSNIA-HERZEGOVINA	NETH.	NETHERLANDS
C.	CROATIA	S.	SLOVENIA
CENT. AF. REP.	CENTRAL AFRICAN REPUBLIC	SL.	SLOVAK REPUBLIC
CZ.	CZECH REPUBLIC	SWITZ.	SWITZERLAND
L.	LIECHTENSTEIN	U.A.E.	UNITED ARAB EMIRATES
LITH.	LITHUANIA	U.S.A.	UNITED STATES OF AMERICA
		Y.	YUGOSLAVIA

Antarctica

Modified Gall Projection

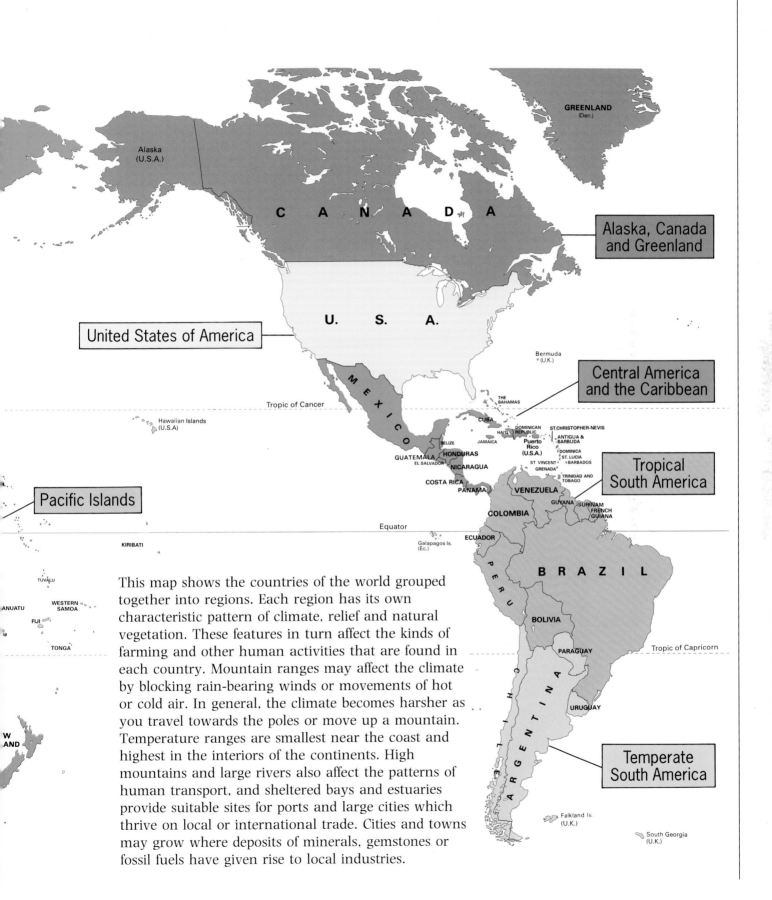

Alaska, Canada and Greenland

United States of America

Central America and the Caribbean

Tropical South America

Pacific Islands

Temperate South America

GREENLAND (Den.)

Alaska (U.S.A.)

C A N A D A

U. S. A.

M E X I C O

Bermuda (U.K.)

Tropic of Cancer

Hawaiian Islands (U.S.A)

THE BAHAMAS

CUBA

DOMINICAN REPUBLIC

HAITI

ST.CHRISTOPHER-NEVIS

ANTIGUA & BARBUDA

BELIZE

JAMAICA

Puerto Rico (U.S.A.)

DOMINICA

ST. LUCIA

GUATEMALA

HONDURAS

EL SALVADOR

NICARAGUA

ST. VINCENT

GRENADA

BARBADOS

TRINIDAD AND TOBAGO

COSTA RICA

PANAMA

VENEZUELA

GUYANA

SURINAM

FRENCH GUIANA

COLOMBIA

Equator

ECUADOR

Galapagos Is. (Ec.)

KIRIBATI

P E R U

B R A Z I L

TUVALU

BOLIVIA

ANUATU

WESTERN SAMOA

FIJI

PARAGUAY

Tropic of Capricorn

TONGA

A R G E N T I N A

URUGUAY

W AND

This map shows the countries of the world grouped together into regions. Each region has its own characteristic pattern of climate, relief and natural vegetation. These features in turn affect the kinds of farming and other human activities that are found in each country. Mountain ranges may affect the climate by blocking rain-bearing winds or movements of hot or cold air. In general, the climate becomes harsher as you travel towards the poles or move up a mountain. Temperature ranges are smallest near the coast and highest in the interiors of the continents. High mountains and large rivers also affect the patterns of human transport, and sheltered bays and estuaries provide suitable sites for ports and large cities which thrive on local or international trade. Cities and towns may grow where deposits of minerals, gemstones or fossil fuels have given rise to local industries.

Falkland Is (U.K.)

South Georgia (U.K.)

Northern Europe

In 1963 a new volcano erupted, and a new island, Surtsey, rose from the sea. Surtsey now stands some 173 m (568 ft) above the sea.

Fjords were carved out by the ice during the last ice age. The longest fjord in Norway is the Sognefjord. It is 204 km (127 miles) long, and at its deepest point is 1,296 m (4,252 ft) deep.

Northern Europe is a region of great contrasts. Within its boundaries are active volcanoes and hot springs, snow-capped mountains and flat lake-studded landscapes, cascading waterfalls and slow meandering rivers. It has a long coastline with steep-sided cliffs and fjords in the north and mudflats, sandy beaches and sand dunes in the south.

The region lies in the westerly wind belt, and has very variable weather. Away from the sea, the climate is more extreme, with hot summers and long, cold winters with a lot of snow. Parts of the north are so cold in winter that the sea freezes over. By contrast the Scilly Isles, off the southwest coast of England, are warmed by the Gulf Stream, and are important producers of early spring vegetables and flowers.

A land carved by ice

In the far north lie the frozen Arctic islands of Svalbard, home to reindeer and polar bears, which cross from island to island in winter when the sea freezes over. The highest part of the region is mountainous Scandinavia. These mountains continue to the west as the highlands of Scotland, North Wales and northern Ireland. This area was covered by ice sheets and glaciers during the recent ice age. They carved out deep U-shaped valleys, many of them now flooded by the sea to form sea lochs (very deep inlets) and steep-sided fjords.

Lowlands and lakelands

East of the Scandinavian highlands are the lowlying lakelands of Finland and the Baltic

▶ Melting snow and glaciers feed Lake Gjende in the Jotunheim mountains of Norway. The lake lies in a valley which has been deepened and widened by a glacier. In the background a glacier flows down from a high snow field. The jagged peaks have been shaped by the action of ice and frost.

countries of Estonia, Latvia and Lithuania, which form part of the North European Plain. The retreating ice deposited vast quantities of sand, gravel and clay here and in the lowlands of the British Isles to form low hills and valleys, through which slow-flowing rivers wind their way to the sea. On the coasts of Denmark, the Netherlands and Belgium, the Plain ends in sandy beaches and mudflats backed by sand dunes.

Land of fire and ice

Iceland forms the highest part of an underwater mountain range, the Mid-Atlantic ridge, that runs almost down the

middle of the Atlantic Ocean. Lava from active volcanoes is still adding to Iceland. The hot rocks and steam provide energy for local industry, and for heating Icelandic homes.

Blessed with natural resources

Northern Europe has a wide range of natural resources. The coniferous forests covering much of Sweden and Finland are exploited for timber, wood pulp and newsprint. The many rivers in the north have been tapped for hydroelectric power. Large fields of oil and natural gas have been discovered beneath the North Sea, and provide valuable export earnings for Britain and Norway. There are large coalfields, too, in the British Isles, the Netherlands and Belgium.

The region also has deposits of minerals such as iron, copper, tin and other valuable metals. Uranium occurs in Estonia. Iron ore deposits in Sweden and the British Isles formed the basis of large-scale heavy industry. Chemical and steel industries have grown up around the old coalfields. Cornwall in southwest England has some of the largest china clay deposits in the world: an important export.

▲ **Finland is a country of lakes and forests. Here, very ancient rocks have been scoured by ice and lakes now fill the hollows. The northern lakes support a rich wildlife in summer, when thousands of migrating birds arrive to breed. Midges, mosquitoes and a host of other insects breed in the lakes and bogs, providing a plentiful food supply for birds to feed to their young.**

◄ **The erupting Krafla volcano in Iceland makes a spectacular display. Red-hot lava welling up along the mid-Atlantic ridge forms a series of 'fire fountains'.**

► Mixed farms are common in the lowlands of Britain, the Netherlands and the Baltic countries. Here, in Hampshire, England, sheep are grazing near fields of rape and wheat.

The Baltic Sea is the largest expanse of brackish (partly salty) water in the world. It is over 1,600 km (1,000 miles) long. Parts of the Sea freeze over in winter.

Reindeer are reared by the Sami people (Lapps) in parts of northern Norway, Sweden and Finland. They provide meat, milk, skins and wool. Their antlers are a source of ivory, which is sometimes carved to sell to tourists.

From woollens to whisky

With no shortage of energy resources, the region has a lot of light industry, producing electrical equipment, shoes, clothing and so on. Well-known examples include glass production in Sweden, Finland and the British Isles, bicycles in the Netherlands and knitwear and tweed in the Scottish islands. England is a centre of printing and publishing. It also once had a thriving textile industry based on local wool and on cotton from America. Today it specializes in high quality and fashion goods. Ireland and Scotland are famous for their whisky.

The fertile lowlands

Britain and the Netherlands, with their small land area and large populations, go in for intensive farming, using lots of fertilizer. Farmers grow fruits, vegetables, sugar beet and cereals (especially wheat) and rear livestock. Meat is exported to other parts of Europe. Mixed farming is also common in Estonia and Lithuania. The Dutch have gained some new land – called polders – by reclaiming it from the sea. The Netherlands are famous for their flowers, especially bulbs such as tulips, which are exported all over the world. Cereals thrive on a lowland belt from eastern Britain to the Ural Mountains in Russia, with wheat on richer soils, and oats and rye on poorer soils and wetter land.

Seafaring nations

The North Sea once had plentiful stocks of commercial fish. However, fish stocks have declined drastically as a result of over-fishing. North European boats also fish much farther afield in the North Atlantic. Until the recent international ban, Norway and Iceland were among the world's leading whaling nations.

The English Channel is an important shipping route, vital to international trade. Important trading ports have grown up around northern Europe's long coastline; Rotterdam, Antwerp, London and Amsterdam are examples. The region has produced many great explorers, and some countries formerly had extensive overseas colonies. This has produced a tradition of overseas trade, merchant shipping and shipbuilding, which remain important today. The recently completed Channel Tunnel provides a rail and motorail link between England and France.

Today tourism is a growing industry. The historical capitals of the region, such as London, Copenhagen and Amsterdam, attract large numbers of visitors, as do the scenic landscapes of Scandinavia and Iceland, the English Lake District, the Scottish Highlands and the lakes and hills of Ireland. Norway attracts winter sports enthusiasts.

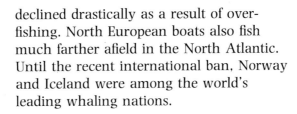

The Netherlands have the longest sea dam in the world. The Afsluitdijk stretches for 32·5 km (20 miles) across the mouth of the Zuider Zee. It is helping to turn the Zuider Zee into a freshwater lake, the Ijsselmeer, whose shores can be gradually reclaimed for agriculture.

◄ The lowlying region of the Netherlands is famous for its bulbs, which flourish in the mild maritime climate. Both bulbs and cut flowers are exported worldwide.

◄ Ancient traditions and ceremonies, palaces and castles, are an important tourist attraction in many parts of northern Europe, especially in England and Scotland. This is the annual ceremony of Trooping the Colour in London, at which the Queen is always present.

Central and eastern Europe

▲ Vineyards are common on warm hill slopes in France, Germany and Luxembourg. Many famous wines are produced in this region and exported all over the world.

► All the family is involved in the potato harvest in the Tatra Mountains. Farming in many parts of Poland is not heavily mechanized.

Europe extends from the Atlantic Ocean to the Black Sea in the east, the Baltic Sea in the north and the Mediterranean Sea to the south. The southern margin is mountainous, with the high peaks of the Pyrenees separated by the French Riviera from the snow-capped ranges of the Alps. In the southeast the great arc of the Carpathian Mountains curves through Romania, then swings northwest towards the Polish border. There are many smaller areas of less rugged high ground, such as the uplands of Bohemia and the Massif Central in France.

Westerly winds blowing from the Atlantic Ocean bring mild, changeable weather to western France and northern Germany, with rain all year round. Summers become hotter and drier and winters colder as you go east. Over much of the region snow falls in winter, especially in the mountains. The south of France has a Mediterranean climate, with hot dry summers and mild winters.

The North European Plain

The land to the north is mostly lowlying, though it is interrupted by highlands such as the mountains of Bohemia and the Black Forest in Germany and the Massif Central in France. The North European Plain extends across Poland and northern Germany to western France. Parts of the Plain are poorly drained and very marshy, especially in Poland and along the German coast.

The Plain was once covered in dense forest, but today most of it is cultivated. The main crops are vegetables, potatoes, sugar beet and cereals. Fruit orchards thrive in hillier

reas, and there are vineyards on warm hill
opes, especially in southern Germany and
rance. In the south, summer fields are
ellow with sunflowers, grown for fodder
nd oil. Warmth-loving crops like melons,
ce and tobacco are grown in Hungary.

igs, cattle and poultry are also reared on
he plains. The mild weather of northwest
rance supports lush pastures for dairy
attle, the source of famous cheeses such as
amembert and Brie.

The wheat belt

Wind-blown soils called loess stretch in a
elt from eastern Britain to the fringes of
he Carpathians. These are fertile soils well
uited to growing wheat. In the far south-
ast, the plains of Hungary (sometimes
alled the *puszta*) and the Romanian
owlands which border the Black Sea are
lso major cereal-producing areas.

Mountain life

The spectacular landscape of the Alps and
Pyrenees attracts sightseers and walkers in
summer and skiers in winter. Many
national parks protect alpine wildlife such
as golden eagles and chamois. The
mountain villages with their pretty wooden
chalets and flower-filled window boxes are
an added tourist attraction. Long tunnels
and remarkable bridges and viaducts carry
roads and railways through the mountains
and across the valleys. The mountain
passes are important routes for commercial
lorries travelling from northern and central
Europe to Italy and Greece.

Cattle graze on the lush alpine pastures, but
spend the winter in the valleys. Their milk
is used in Swiss chocolate and famous
cheeses such as Gruyere and Emmental. In
the valleys, farmers grow hay for winter
fodder.

▲ Glaciers are still
carving out valleys in
the Alps. The rugged
terrain presents an
awesome challenge for
mountaineers and
alpine skiers.

Bison, elk, wild boar,
wolves and bears still
roam the Bielowieza
Forest in Poland. This is
one of the few remnants of
the forests that once
covered the North
European Plain.

▶ A steelworks at Katowica, in Poland. The former communist countries of Eastern Europe have a lot of heavy industry which has caused a great deal of pollution.

Poland's national drink, vodka, is made from cereal grains or potatoes. France is famous for its wines, and Germany for its beers, brewed from hops grown in the southern valleys.

In the mountains of the Pyrenees and the Alps and Bohemia most people live in wooden chalets. The roofs have large sloping eaves so the snow can slide off.

▼ The River Danube sweeps through Budapest, the capital of Hungary. The Danube is an important waterway for transporting goods between several European countries.

Swiss clocks and space rockets

The region has some major coalfields, including Germany's Ruhr district and Silesia in Poland. Romania has oil and natural gas, and the many rivers of the region are used for hydroelectric power. France depends heavily on nuclear power stations. Together with iron and other metal ores, these resources provide a sound basis for industry, especially engineering and manufacturing based on iron and steel.

Poland has important shipyards; cars and trucks are produced in France, Germany, the Slovak Republic and the Czech Republic and machinery in Hungary and Romania. There are also specialist industries such as optical instruments and electronic goods. Neuchatel in Switzerland is the centre of clock and watch making. France has many science-based industries, famous for the Concorde and Airbus aircraft, Ariane space rockets and high speed trains.

The region is paying a price for its many industries as there is a serious pollution problem. Acid rain is killing the forests of Germany and Poland, and even trees in the Alps are suffering from traffic fumes. In the cities children's health has suffered.

Many great rivers flow from the mountains to the plains. Waterways like the Rhine, Meuse, Elbe, Seine and Danube are important for transporting raw materials and manufactured goods, coal and other cargoes.

The varied landscapes of Central Europe, together with its great cultural heritage of castles, palaces, opera houses, fine cities and picturesque villages, attract many tourists.

Southern Europe

outhern Europe has a long coastline on the Mediterranean Sea, with many islands, peninsulas and bays. The Mediterranean Sea is linked to the Atlantic Ocean on the west through the narrow Strait of Gibraltar at the foot of the Iberian Peninsula (Spain and Portugal). To the east lies the inland Black Sea. The northern boundary is mountainous: the peaks of the Pyrenees rise to 3,400 m (11,150 ft), and the Alps reach above 4,500 m (15,000 ft); the mountains of the Balkans are lower. Much of the Iberian Peninsula is a large plateau broken up by deep river valleys. The plateau slopes roughly from west to east, and the high land keeps out rain-bearing winds from the Atlantic. Here, summers are very hot and winters very cold.

Holiday weather

The Mediterranean has a climate almost tailor-made for holiday-makers. Winters are mild and wet, but summers are hot, dry and sunny. Rainfall decreases eastwards with distance from the Atlantic Ocean, especially in areas such as Greece and the interior of the Iberian peninsula, which are shielded from the westerly winds by mountains. Here, winters are more severe, too. The mountains have lower temperatures, higher rainfall and lots of snow in winter.

▼ In the great city of Istanbul, Turkey, the cultures of Europe and the Middle East meet. Business-men and women in western dress mingle with peasants in traditional clothing. The minarets of mosques and the spires of Orthodox Christian churches mingle on the city skyline.

Olive groves dominate the dry Spanish landscape. Olives are grown throughout the Mediterranean region. They are an important part of the local diet. Olive oil is used for cooking and is exported to many other parts of the world. Olive wood is often carved into souvenirs for tourists.

► **This picturesque harbour on the Greek island of Hydra is typical of the island scenery that attracts tourists to the Mediterranean coast. Local fishermen still supply the many hotels and tavernas with fresh fish and shellfish.**

The Highway of the Sun is a motorway that runs down the spine of Italy, passing through many tunnels. It is one of the great scenic routes of Europe.

The flat roofs of houses on the Mediterranean islands are used for drying fruit in summer and collecting rainwater in winter.

The Corinth Canal, which links the Gulf of Corinth to the Aegean Sea in Greece, is the deepest cut ever made. It is 6·3 km (3·9 miles) long, 8 m (26ft) deep, with a maximum depth of 459 m (1,505 ft). Passenger ships are pulled through by two tugs.

Small-scale farming

The forest that once covered much of the land was cleared thousands of years ago. In places too dry to be farmed, the natural vegetation is thorny shrubs and trees with leathery leaves, and aromatic herbs such as marjoram and thyme. Stone-built villages cling to steep hillsides and farmers have to work hard make a living from the stony soil. Hillside terraces trap valuable water for olive and fruit trees.

Most farms are small. Farmers grow wheat, vegetables and flowers, and rear sheep and goats for cheese and meat, mostly for local consumption. In Turkey, Angora goats are reared for their soft wool. Terraced groves of olives with twisted trees separated by low walls are a familiar sight. Olives are eaten and their oil is widely used for cooking. Grape vines also do well here, and Spain, Portugal, Italy and Bulgaria are exporters of wine.

Other crops include maize (grown for fodder), rice on the floodplain of the River Po in northern Italy, cotton and tobacco in Greece, Bulgaria and Turkey, and fruits such as almonds, apricots, oranges and lemons. There are cork oak plantations on the dry, high plateau of the Spanish Meseta. In the mountain valleys, hay is grown for winter fodder. Cattle are taken to alpine pastures to graze in summer, and return to the lowlands in winter. In recent years many people have left this hard life for an easier one in the cities.

Living off the sea

Many people make a living from the sea, selling fish and shellfish, as they have for thousands of years. Ancient fortresses still guard small rocky harbours. Today giant tankers bring oil through the Suez Canal, and luxury liners cruise through the islands. Modern yachts mingle with traditional fishing boats. But catches of anchovies, sardines and tunny are falling as a result of pollution, which is becoming a serious problem in the Mediterranean. Growing coastal cities pour sewage into the sea, and oil pollution is common. There are hardly any tides here, so the shores are not flushed clean by the sea.

Exporting sunshine

Tourism is a major source of income for the whole region. Millions of tourists visit the Mediterranean beaches, especially during the long, sunny summer. In winter, skiers flock to high mountain resorts. Sightseers are drawn to the wild snow-clad Alps and Pyrenees, the wildlife of national parks such as Gran Paradiso and Abruzzi in Italy, and the picturesque villages. Cities like Venice, Rome, Athens, Barcelona and Dubrovnik offer the tourist a feast of art and culture.

Hydroelectricity from the region's many mountain rivers supports various manufacturing industries in the foothills and on the plains. Albania has petroleum and natural gas reserves, and Bulgaria has coal deposits. There are scattered metal ores in the region. Specialist industries include cars, made in Spain and Italy. Racing cars are tested on the Monza circuit near Milan. Spain also has a shipbuilding industry on the Atlantic Coast. Italy has a textile industry based on synthetic fibres, and is an exporter of shoes and fashion goods.

◄ **Many Bulgarians and foreign tourists enjoy holidays at beach resorts on the Black Sea. This one is in Sonnenstrand.**

Russia and its neighbours

▲ **There are many active volcanoes in the Kamchatka Peninsula, in the far east of Russia.**

300 rivers flow into Lake Baykal, the world's deepest freshwater lake. Only one river, the Angara River, leaves the lake.

The river Ob and its tributary, the Irtysh, form the world's fifth longest river. From the source of the Irtysh in northern China to the mouth of the Ob in the Arctic Ocean is a distance of 5,570 km.

Russia is a vast country that stretches around the roof of the world from the Baltic Sea in the west to the Pacific Ocean in the east. Its northern boundary is the Arctic Ocean. Its southern limit are the steppes of Kazakhstan and the great mountain ranges on the border with China. The Ural Mountains divide the country into a western and an eastern section. Over 9,000 km (5,600 miles) from east to west, Russia crosses 11 time zones. While people are eating their breakfast in the European part, they are sitting down to dinner at the end of the same day, in the Asian far east.

A treeless wilderness

In the far north the coasts and sea are ice-bound in winter, and the land is a 'polar desert', with very little plant cover. Travelling south, this merges with the tundra, a huge treeless expanse covered in low-growing cushion-shaped plants, mosses and lichens. The ground here is frozen all year round, except for the surface layers of the soil, so there is not enough free moisture for trees to grow. Frozen ground tends to bulge after it is disturbed. In some Siberian cemeteries coffins buried in summer are pushed to the surface again the following winter. In summer the tundra changes into a glittering landscape of pools, marshes and meltwater streams, a breeding ground for midges, mosquitoes and countless other insects. These attract huge numbers of migrating birds, especially ducks, geese and waders, who come to breed here.

From forest to fire

Farther south again is the taiga, the world's largest coniferous forest. Some of the world's longest rivers – the Irtysh, the Ob, and the Yenisey – cross this region. The taiga is the world's largest forest resource. It is increasingly being exploited by foreign companies. Timber is floated down the rivers to sawmills, factories and ports.

Farther south there is again a lack of water, this time due to low rainfall. This affects the tree cover, and the forest gives way to grassy steppes. Along the border with China is a series of high snow-capped mountain ranges, part of the great Himalayan system. Between the ranges, high basins, cut off from rain-bearing winds, form cold deserts. Rivers rising in these mountains feed the world's largest freshwater lake, Lake Baykal, which in places is 1,741 m (5,712 ft) deep. The lake used to be famous for its crystal clear waters but it is now badly polluted with chemicals from pulp and paper mills and

▲ A winter scene in a Siberian town. The temperature stays below feezing for months on end.

her industries around its shores. Lake ~~~ykal is home to the rare Baykal seal, one ~the few seal species to live in fresh water.

~the far east, the Kamchatka Peninsula is ~mountainous region containing about ~0 volcanoes, 67 of them still active. They ~rm part of the 'Ring of Fire': volcanoes ~rown up around the Pacific Ocean where the great crustal plates of Asia and the Pacific Ocean floor are colliding. The underground heat is used to heat houses and generate electricity.

The great divide

The north-south sweep of the Ural Mountains separates European Russia from Asian Russia. The mountains continue northwards as the Novaya Zemlya islands. At their highest point, Mt Narodnaya, the Urals are less than 2,000 m (6,500 ft) high. To the west of the Urals, the climate is generally moist and not too severe, but to the east it becomes progressively drier, with bitterly cold winters and short, hot summers.

The partly forested Urals have been extremely important throughout the many centuries of Russia's history because they have great mineral wealth. Dense populations are centred around industries based on deposits of iron, copper, nickel, gold, platinum, asbestos and gemstones. There are also valuable energy resources here in the form of oil, coal and lignite.

◄ In parts of Siberia, reindeer herding is a traditional way of life. The reindeer graze on the tundra in summer,and are led to the edges of the forest (the taiga) in winter. The animals supply meat, hides and also their antlers, which are shed each year. In the past, whole families used to travel with the reindeer, but today the herders are mostly men. The women and children stay in the villages, where the schools are.

From snowy plains to sunny beaches

Mt Elbrus, in the Caucasus Mountains, is Europe's highest peak. This extinct volcano rises to 5,642 m (18,525 ft) above sea level.

▼ **The River Moskva winds its way through Moscow, the capital of Russia. The picturesque old walls of the Kremlin in the foreground contrast with the severe tower blocks of flats typical of the communist era.**

In the northwest the Volga basin lies in the heart of the great North European Plain, snow-covered in winter and marshy in summer. The Pripet marshes in Belarus and Ukraine form the largest swamp in Europe. Many rivers cross the southwest on their way to the Black Sea. Farming here is intensive, growing wheat, sugar beet, sunflower seeds (for oil), flax, potatoes, vegetables and melons. The Plain is also a major livestock-rearing area.

Georgia is a mountainous country, with many snow-capped peaks over 4,600 m (15,000 ft). The highest peaks have tundra vegetation. Further down the slopes, alpine meadows grade into conifer forests, then broad-leafed forests with wild fruit trees. Lowland Georgia is sheltered from the northerly winds by the Caucasus Mountains, and has a subtropical climate. Around the Black Sea, farmers grow fruit

trees and commercial crops such as hemp, tobacco, tea and flax. This is also a major wine-producing area. The Black Sea coast popular with tourists, who flock to the many resorts and health spas.

locomotives, airliners and even ships. There are also lighter manufacturing industries based on chemicals and textiles.

A vast granary

The Russian steppes cover an area of 3·5 sq km (1·4 million sq miles). Much of the area is covered in a dusty wind-blown soil called loess, which is very fertile. It is here that most of the region's cereal crops are grown, mainly wheat, with lesser amounts of barley, rye, millet and maize. Cattle are raised for meat and milk, and pigs are also kept. Sheep are more important in western Siberia, and reindeer in the far north. Wind erosion is a problem on the steppes. Farmers build wind-breaks around their fields. They store water in ponds and behind dams in sheltered ravines.

◀ A factory in Knovokuzentsk, a small mining town in Siberia. Russia and neighbouring countries are rich in mineral resources and fossil fuels, which support a wide range of manufacturing industries.

▼ In parts of Georgia, sheltered from the icy north winds by the Caucasus Mountains, the climate is warm enough for tea-growing. The special tractors are harvesting the tea leaves.

Blessed with resources

The region contains large deposits of almost all the minerals used in modern industry. It also has plenty of coal, oil, natural gas and rivers to generate hydroelectric power. Siberia is especially rich in resources, but winters there are bitterly cold. The government pays extra money to encourage people to move to these areas to work in the mines, because living conditions are not good here. In the north the Sun may shine for only about an hour a day for months. Sunshine gives important vitamins to the body, so children have to stand around ultraviolet lights for a certain time each day at school to make up for the lack of sunlight. The Russians are now experimenting with huge mirrors in space, which they hope may one day light up parts of Siberia in winter. The cold makes the mining difficult. Special machinery is needed to work in freezing conditions and all buildings must be heavily insulated to keep them warm.

Elsewhere, industry is particularly concentrated in Belarus, Georgia, Ukraine and the Caucasus Mountains. Engineering industries are important, producing machine tools, engineering equipment,

Central Asia

▲ The fertile Bamiyan Valley in Afghanistan is a stark contrast to the surrounding desert. Much of Central Asia is arid. Melting snow from the high mountains feeds many rivers, which provide water for irrigation. A wide variety of fruits and cereals are grown in the sheltered valleys.

Central Asia is an inland region of deserts, steppes and high mountains. In the west are two great lakes, the Caspian Sea and the smaller Aral Sea. Across the north lie the vast steppe grasslands of Kazakhstan, and in the drier south are the great deserts of the Karakum and Kyzylkum. A series of high mountain ranges sweeps in a huge arc across the southeast – the Hindu Kush in Afghanistan, the Pamirs in Tadzikistan and the mighty Tien Shan range in Kirghistan. In the southwest the woodlands, meadows and steppes of Azerbaijan and Armenia clothe the warm southern flanks of the Caucasus mountains.

A land of extremes

Being in the heart of the continent, far from the ocean, the region has very hot summers and very cold winters. Summer temperatures in the Karakum Desert can reach 50°C (122°F) in the shade, yet in winter the temperature can plunge to -33°C (-27°F) as cold air pushes south from the Arctic. In the southwest, the Caucasus Mountains shield the lowlands of Armenia and Azerbaijan from these northern air masses. Here, the climate is ideal for grain crops and for fruits such as peaches, grapes, pomegranates and figs.

The mountains keep out moisture-bearing winds, so rainfall and snowfall are low, specially in the south. Many rivers peter out in the desert as their waters evaporate in the summer heat. But the region does contain one of the world's largest fresh-water lakes, Lake Balkhash.

Once nomads roamed the steppes and semi-deserts with their sheep, goats and camels. Today the farmers have settled in villages and towns, but in the mountains people still move with their animals to alpine pastures in summer, and back to the shelter of the valleys in winter.

The sea that disappeared

Irrigation is increasing the amount of land that can be cultivated. But this is not always successful. So much water has been taken from the rivers feeding the Aral Sea, that the sea is drying up. Important fishing industries have been lost, and empty 'ghost' villages and stranded ships line the Aral Sea's former shores. Dried out salt and pesticides from the old sea blow for miles, poisoning the soil and the water supplies.

From cereals to carpets

Cereals are the basic food crop, mainly wheat, barley and maize. Cash crops are also important. Azerbaijan and Uzbekistan are major cotton producers, silkworms are reared for silk in Tadzikistan and

Uzbekistan, and Afghanistan's Karakul sheep provide fine wool, so textile industries are common. Local carpets are world-famous. Tobacco and wine grapes are grown in the Caucasus, geranium oil (for perfume) in Tadzikistan, and opium poppies wherever they will grow.

The region has a great wealth of natural resources. Oil and natural gas are found in Azerbaijan, Tadzikistan, Turkmenistan and Uzbekistan, and coal deposits are wide-spread. The rivers rushing down the mountains provide hydroelectric power for industry. Deposits of phosphates and other salts, and metal ores of iron, copper and other metals form the basis of manufac-turing, engineering and chemical industries.

◄ Melon-sellers are doing a good trade in the market at Samarkand, in Uzbekistan. The juicy fruits are much sought after in the heat of summer. Most of the women here are wearing traditional Muslim dress. Some of the younger people, however, are in western clothes.

The highest peak in Central Asia is Mount Communism in the Pamirs, 7,495 m above sea level.

The world's longest irrigation canal, the Karakumsky Kanal, stretches for 1,200 km (736 miles) across the Karakum desert in Turkmenistan. About two thirds of the canal is wide and deep enough to allow ships to use it.

Lake Balkhash is 605 km (376 miles) long. It is one of the world's largest freshwater lakes.

Steppe grasslands once covered most of the northern part of Central Asia, but they have largely been ploughed up to grow cereals, and the region is a major wheat exporter.

Central Asia is rich in metal deposits. Ores of copper, iron, lead, zinc, gold, silver, vanadium, tungsten, tin, chromium, nickel, cobalt, antimony and mercury form the basis of industries which produce many goods for export.

◄ Oil rigs fill the landscape at Baku, in Azerbaijan. There are still large undeveloped reserves of oil and natural gas in Central Asia.

Middle East

The Middle East is a region of vast deserts and high arid mountains. But it includes the region of Mesopotamia, between the great Euphrates and Tigris rivers in Iraq, sometimes called 'the Cradle of Civilization', the site of the world's oldest civilization. Three of the world's greatest religions, Judaism, Christianity and Islam, arose in the Middle East. The region has long been a crossroads for traders and armies, linking the continents of Europe, Asia and Africa.

▼ In the vast swamps of southern Iraq, reeds are used for many different purposes. This boat-load will be used to make walls and mats for this Iraqi island house.

Arabia and beyond

The great plateau of Arabia dominates the region, tilting gently eastwards from the mountains along its western edge. They are deeply dissected by steep canyon-like valleys called wadis, carved out by the flash floods that occur after sudden desert storms. East of the mountains is a series of deserts. The Rub Al Khali (the Empty Quarter) in the south is the largest uninterrupted sand

desert in the world. It covers about 595,700 sq km (230,000 sq miles), and in places the dunes are over 150 m (500 feet) high. The coastal plains are widest on the Gulf coast. The Arabian coast is fringed by some of the world's most spectacular coral reefs.

The lowland desert and steppe grasslands of Iraq merge into green and cultivated countryside around the Tigris and Euphrates rivers. Its neighbour, Iran, is a high country with towering snow-capped mountains in the north. More than half the country is desert, covered in thick crusts of salt.

The climate is harsh. Summers are hot and very humid by the Gulf. Temperatures can soar to 54°C (130°F), yet snow may fall on the mountains in winter, especially further north, where winters can be severe. The Arab peoples of the region wear mostly long loose clothes to protect themselves from the fierce summer Sun. In the south-west, monsoons from the Indian Ocean bring more rain. Bedouin nomads have roamed the region for thousands of years with their herds of goats and sheep, moving from well to well, but most people live in villages and towns.

Riches in the desert

The Red Sea and the Persian Gulf and their coral reefs are rich sources of fish and other seafood. Date palms grow wherever there is enough moisture, and dates form the staple diet in Arabia. Alfalfa is often planted between the palms as a fodder crop. Wheat, barley and millet are grown on the irrigated plains and in moist wadi beds. Cotton and rice thrive where there is enough water. Many different fruits grow here, including melons, pomegranates, mangoes, figs, grapes, bananas and citrus fruits. Tobacco is grown as a cash crop. Cattle and poultry are reared near the towns, and camels and horses are bred for racing.

he region's real wealth comes from oil. etroleum and natural gas have been found many parts of the region. Petrochemical dustries are important in many countries. ahrain has become the financial centre of e Middle East. Saudi Arabia also has xtensive mineral ores.

Greener lands

he lands fringing the Mediterranean Sea ave a better supply of rainfall. The ountains of Lebanon and Turkey are artly forested; the cedars of Lebanon are orld-famous. Israel in particular makes reat use of irrigation. It is one of the orld's leading producers of oranges and rapefruits. It also produces early vegetables for the European market, apples, cotton, wheat and grapes. Sheep, cattle and turkeys are reared, and fish are farmed. Turkey exports tobacco, cotton, high quality grapes for drying, olives and figs, wool from angora goats, carpets and textiles. The interior Anatolian plateau and the Syrian Desert, once covered in steppe grasslands, are important wheat-growing areas. Mineral resources support manufacturing industries, but its main industry is textiles.

Tourism is an important source of income. The Mediterranean climate, fine beaches and coral reefs, the region's religious and archaeological history, and ancient cities like Jerusalem and Istanbul are an exciting holiday mix.

There is fresh water deep beneath the desert. In some places it comes close to the surface at oases, or bubbles up as springs, but it can also be reached by wells and pumps.

Another source of water is the sea. Expensive desalination plants take out the salt to produce fresh water.

Favourite desert pastimes include hunting with falcons, and racing camels and horses.

Turkey's highest peak, Mount Ararat, at 5,165 m (16,945 ft), is suggested as the resting place of Noah's Ark.

Iran is famous for its hand-made carpets (Persian carpets) among the most valuable in the world. Iran is the largest non-Arab country in the Middle East.

◀ A green field of young wheat between desert sand and rocky outcrops in Saudia Arabia. Irrigation is necessary for successful farming in most parts of the Middle East.

Indian subcontinent

The very high part of Nepal near the Chinese border is sometimes called 'the Roof of the World'. It contains the world's highest mountain, Mount Everest, whose summit is 8,863 m (29,078 ft) above sea level.

Cox's Bazar has the world's longest beach, 121 km (75 miles) long.

Cotton was an important crop in Bangladesh in the 18th century. The cotton known as muslin was named after the Muslim cotton weavers of the capital Dhaka.

The Indian subcontinent contains some of the most dramatic and varied scenery on Earth, ranging from the snow-capped peaks and glaciers of the world's highest mountain ranges to the mangrove swamps of the Sundarbans, the tropical jungles of Assam and the Western Ghats, and the sandy wastes of the Thar Desert.

The monsoons bring torrential rain between May and October. In the lowlands the weather is hot for most of the year. Before the monsoon the heat becomes almost unbearable – the temperature can reach 48°C (120°F). During the hottest months of the year in the lowlands most people try to stay indoors, or in the shade by day. At night they may sleep on the roof to get cool. The monsoon rains can cause devastating floods as the rivers overflow their banks. In autumn, tropical storms called cyclones sometimes cross the coastal areas. In November 1970 a cyclone killed one million people in Bangladesh. Bangladesh gets 2,000 mm (80 in) of rainfall a year, almost all of it during the monsoon

The great deltas

Most of the people live in the great river valleys and on the deltas. In Bangladesh the floodplains of the Ganges, Brahmaputra and Meghna rivers merge to form the largest delta in the world. Here, it is easier to travel by boat than by road. This is one of the most densely populated regions on Earth. Rich sediments deposited by the rivers make the delta very fertile. Irrigation canals direct water into small fields rimmed by mud walls, and the fertile soil supports a great variety of tropical and subtropical crops. Water buffalo are used to pull carts and plough, and to pump water. Planting, weeding and harvesting are done by hand.

Rice is the staple food crop in the hot steamy lowlands, and wheat in the cooler parts of India and the Indus valley. Lentils, root crops and fruits, such as melons, pineapples, bananas and mangoes are grown. Cash crops include sugar cane, oilseeds, coffee, tobacco, sesame, groundnuts (peanuts) and rubber. The main export crops are jute (used for making rope and

▶ **Water buffalo and farmer in the fertile rice paddies of Bangladesh**

acks), cotton, and tea, which is also grown
1 the hills. On the higher ground inland
re valuable teak forests.

heep, goats and cattle are also reared on
he plains. Bangladesh exports hides and
eather goods. India has the world's largest
opulation of cattle, but most are
ndernourished and do not produce much
1ilk. Cattle are considered sacred in many
arts of India, and it is forbidden to kill
hem. Fishing is important around the
oast. Tilapia and carp are reared in inland
sh ponds.

Jatural gas, petroleum, coal, hydroelectric
ower and mineral reserves help support a
vide range of industries, from textiles and
ottery to bicycles and railway engines.

Life in the hills

emperatures decrease as you go up the
nountains. The vegetation changes from
ropical jungle to Tarai grasslands, then to
lpine vegetation. Above about 4,870 m
16,000 ft) the temperature never rises
above freezing, and there is permanent
now and ice.

tice is grown in clearings in the Tarai
grassland, in the jungle-covered foothills
and along valley bottoms, and wheat, millet
and maize are planted on the terraced
ower slopes. The jungles of the foothills
and the plateaus still contain tigers, deer,
vild pheasants and other animals, but their
numbers are dwindling as the growing
oopulations invade the forest margins.

Yaks graze on the alpine pastures in
ummer. Their long woolly coats keep out
he cold. Tourism is important in Nepal, but
t brings problems of waste disposal, litter
and loss of forests for firewood.

The Indian Ocean coast

The great deltas of Bangladesh and India
end in a fringe of dense mangrove swamps,
the Sundarbans. The tangle of mangrove
roots are invaded by narrow fingers of
ocean. The mangrove swamps are
important natural nurseries for many

commercial fish and shellfish which live in
the open ocean as adults. They also form
an important barrier against the sea and
against storms rushing in from the ocean.
In some places mangroves are being cut
down to make way for fish and shellfish
farms, new settlements and even just for
wood chips. This is risky in an area so
exposed to the sea.

Where there is shallow water offshore,
coral reefs are common. Shallow lagoons
between the reefs and the shore are home
to fish and shellfish which are easily
collected by local villagers. Many Indian
Ocean islands, such as the Maldives, are
really atolls – ring-shaped islands formed
from coral reefs growing on the summits of
underwater mountains. These islands are so
low that a very small rise in sea level
would completely submerge them.

▲ Hindus visit the holy
River Ganges at
Varanasi, in India.

Far East

▲ **The steppes stretch across Central Asia from Russia and Kazakhstan to Mongolia and China. The yurts (round tents) in this photograph are the homes of nomadic people who live on these vast plains.**

Lowland China was once covered in forest, but this was cleared long ago for agriculture. Soil erosion is a big problem. The Huang He is the world's muddiest river. Its name means 'Yellow River'. In 1931 it caused the worst floods in history, and 3,500,000 people were killed.

The whole region is prone to earthquakes. The worst one in recent times was near the town of Tangshan in July 1976. More than 240,000 people were killed.

The Far East is dominated by China, a huge country of over one billion people, about one fifth of the world's population. In the northeast is a series of high mountain ranges and cold, dry plateaus. The great Gobi Desert extends far into Mongolia, and in the east are the Tibetan Plateau, and the Tarim Basin, which contains the world's driest desert, the Taklamakan. Here, winter temperatures can plunge to -40°C (-40°F).

Groups of nomads live on the high plateaus in tents called yurts, made of animal hair. They raise sheep on the grassy steppes. The yurts are taken down from time to time and moved on horses or camels to new pastures. However, most people now live in villages. The shaggy-coated yak is ideally suited to the harsh conditions of the Tibetan Plateau. It provides milk, meat and transport. Tibetans drink yak butter in tea with salt. North-east China is an important wheat-growing area, and the bamboo forests of Szechuan are home to the endangered panda.

The cultivated lowlands

Much of lowland China is covered in a soft yellow dusty soil called loess, blown south from the Gobi Desert. Loess is very fertile, and most Chinese people live and work in the countryside. The monsoons bring heavy rain in summer, and this is trapped in narrow terraces on the hill slopes. In south China the climate is very warm; tropical rainforests and mangrove swamps fringe the South China Sea. Fishing is important throughout the region. Fish and shellfish are also farmed in small ponds.

When the rivers flood after the monsoon, the water is channelled into paddy fields. Throughout the Far East rice is the main crop. Maize, wheat, sugar cane, sweet potatoes, soybeans, potatoes, cotton and tea are also grown. All kinds of livestock and poultry are reared. In the Chinese country-side asses and donkeys are used to draw carts and carry loads. Bicycles are popular everywhere.

High-tech industries

China and Korea have large reserves of minerals and precious metals. They also have ample supplies of hydroelectric power, coal, petroleum and natural gas. These support expanding manufacturing and engineering industries. By contrast, Taiwan's thriving industries depend on imported raw materials. Like Japan and South Korea, Taiwan exports clothing and has high-tech industries based on electrical goods and electronics. Japan and Korea are also leading shipbuilding nations and car exporters.

The tiny British colony of Hong Kong, on the south-east coast of China, is a major trading and financial centre. It is due to return to Chinese rule in 1997. Hong Kong is one of the most densely populated places on earth, with up to 100,000 people per square kilometre in some areas. The city streets are packed with cars, trucks,

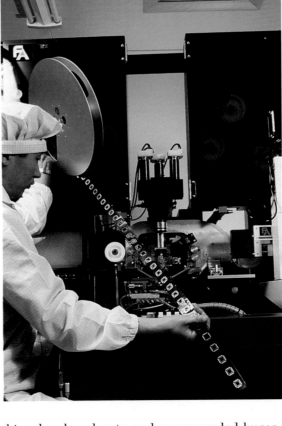

◄ A Japanese worker examines newly-produced silicon chips in an electronics research centre.

bicycles, handcarts and overcrowded buses. The mainly Chinese population uses raw materials from China to produce clothing, electronic goods, watches and clocks, textiles and toys. The port of Hong Kong handles more trade than China itself.

Surrounded by sea

The islands of Japan are the peaks of a partly submerged chain of volcanoes, many of them still active. The climate varies from cold with snowy winters in the north to almost tropical in the Ryukyu Islands. The population is densely packed into the coastal plains. Japan leads the world in building cars, ships and TV sets. It exports a great deal of electronic equipment, such as computers. However, it has to import nearly all its raw materials, as well as coal and oil. Fish rather than meat is the staple diet in Japan. Japan's fishing industry is the largest in the world. Each island also has a large area of farmland, divided up into family farms. Several generations of the family live on the same farm. Rice is the main crop.

The Great Wall of China was built in 221–210 BC and rebuilt later. It stretches for 2,700 km (1,700 miles), and has 25,000 sentry towers. In places it is 10 m (30 ft) thick. It once marked the frontier between the settled Chinese farmers and the nomads who roamed the dry northern lands.

China is trying to reduce its huge population so that people can have a better standard of living. The government is encouraging couples to have only one child.

◄ A landscape in southern China.

South-east Asia

South-east Asia is a vast region, straddling the Equator between the Pacific and Indian Oceans. Only one fifth of the total area is land. This consists of mainland in the north, and the world's biggest archipelago (group of islands) in the south and east. Indonesia alone has over 13,000 islands, only 3,000 of them inhabited. The Philippines is made up of 7,107 islands. The mainland is crossed by large rivers such as the Irrawaddy, Salween and Mekong, which are fed by snowfall in the Himalayas. They wind through the foothills and across vast floodplains to form deltas, pushing out into the surrounding ocean.

▼ Juara beach on Tioman Island, Malaysia. Southeast Asia has thousands of islands surrounded by warm, tropical seas.

Mangrove swamps and tidal waves

Many of the islands are formed from volcanoes that rise out of the sea. Well over a hundred are still active, and earthquakes and tidal waves are common throughout the region. Some of these mountains rise above 4,000 m (13,000 ft). There are thousands of miles of coastline, and in most countries fish are a more important source of protein than meat. Vietnam has

► A floating market in Thailand.

attractive sandy beaches, but much of the region is fringed with mangrove swamps. These are often felled to make charcoal, or to make way for prawn farms and other agriculture.

Land of the two monsoons

Most of the region is warm and wet all year round. The further you go from the Equator, the more seasonal the weather becomes: summers get hotter and winters colder, and there are distinct dry and rainy seasons.

There are two monsoon seasons. From November to February the northern monsoon blows out from the cold air masses over Siberia and China. This brings rain mainly to the south, where the winds have crossed the sea. From April to September the southwest monsoon blows from Australia, bringing rain to the whole region. This is the main rainy season in the north. The central part of the region has very high rainfall, from 2,000-4,000 mm (80-120 in) a year.

Plantations and paddyfields

South-east Asia was once heavily forested, but today much of the land is a patchwork of fields and natural vegetation. Most people work on the land as peasant farmers. Rice is the main crop, grown in mud-rimmed paddyfields that rely on the monsoon rains and river floods to fill them with water. Some of the hill terraces are thousands of years old. In Cambodia farmers pedal rohats, giant wheels which scoop water up and tip it onto the crops. In the floodplains of rivers such as the Mekong and Irrawaddy, a network of canals feeds the paddies.

Other important food crops include maize, cassava, pulses, sweet potatoes and pineapples. Besides sea fishing, carp and tilapia are reared in inland ponds and canals. There are also plantations of rubber trees, sugar cane, bananas and coconut palms (which produce coconuts, palm oil and copra). These are important export crops. Cash crops such as cotton, jute, tea, coffee, tobacco, spices and, in Laos, opium are also grown for export. Myanmar is a major exporter of rice, while rubber is particularly important in Malaysia, Indonesia and Thailand.

◀ Rice terraces in the Philippines. The low walls hold back the water when the terraces are flooded by the monsoon rains.

From timber to tourism

Some of the world's greatest tropical rainforests grow in south-east Asia. They contain many rare animals and plants, including orang utans and priceless orchids. But the rainforests are being cut down at an ever-increasing rate. In hilly areas that have lost their forest cover, the heavy monsoon rain soon washes away the soil, causing mudslides and erosion. Forest peoples lose their way of life as the forest disappears. Timber is an important export, especially in Indonesia, the Philippines and Borneo. In drier inland areas teak is extracted from forests and plantations for export.

Rural economies

Most of the countries in south-east Asia have economies based on agriculture and forestry. Few have significant mineral and energy resources. Malaysia, Myanmar, Vietnam and the Philippines have some oil and natural gas, and Malaysia is is a petroleum exporter. Tin is exported from Malaysia, Laos and Thailand, and Vietnam produces apatite, an important source of phosphate for fertilizers and chemicals.

Singapore is a small but very wealthy country. Once a British naval base, its port and airport are among the busiest in the world. Its thriving light industries include textiles, electronic goods and high quality printing.

The high mountains of Borneo contain vast cave systems. The Sarawak chamber in the Gunung Mulu National Park is the world's largest cave chamber. It is 700 m (2,300 ft) long and on average 300 m (980 ft) wide, with a minimum height of 70 m (230 ft).

The river that flows backwards: the Mekong river in Cambodia flows backwards for about 100 days a year, as a tidal bore from the sea pushes the water upstream.

Borneo is one of the world's largest islands. It covers an area of 751,900 sq km (290,300 sq miles).

Deep in the jungles of Cambodia lie the magnificent temples of Angkor Wat, built by the Khmer people in the twelfth century.

Australia, New Zealand and New Guinea

The people who live in the outback have to rely on flying doctors (doctors in aeroplanes) for medical help.

The native people of Australia are called Aborigines. The name 'Aborigine' comes from the Latin *ab origine* which means 'from the beginning'. The Aborigines have lived in Australia for a very long time.

The Great Barrier Reef, the world's largest coral reef, lies off the east coast of Australia. It stretches for 2,027 km (1,260 miles), and is large enough to be seen from the Moon.

Australia is one of the flattest countries on Earth. Its mountains are very ancient. They have been worn down by wind and water over millions of years, and are now rounded rather than steep. The highest ranges, in the east, have snow in winter. By contrast, New Guinea is a mountainous, thickly forested island. New Zealand has active volcanoes and geysers, snow-capped mountains and fjords.

Very special wildlife

Many Australian animals and plants are found nowhere else in the world. Kangaroos, koalas, wombats, possums and wallabies are all marsupials, which are mammals that keep their young in pouches. The most famous native Australian trees are the eucalypts, which can be found from the deserts to the mountain snows.

New Guinea has many marsupials, but it also has species more typical of Asia. These reached the island at a time when the sea level was lower, or floated there on rafts of vegetation. New Zealand became separate islands before mammals had evolved on nearby continents. The only native mammals here are bats. Many strange flightless birds, such as kiwis, survived here because there were no mammals to hunt them on the ground.

The outback

Australians call the vast inland region 'the outback'. It is hot and dry and a large part of it is desert. The first European explorers, dismayed by endless barren plains, called the central interior 'the dead heart'. However, this was misleading; even the deserts are full of life. After heavy rain, many species of flowering plants bloom,

▶ **Ayers Rock, near the centre of Australia, is sacred to the Aborigines (native people) who call it Uluru. The whites have now adopted 'Uluru' as the name of the rock. It is 348 m high and over 2 km long. At sunset the sandstone glows a deep red.**

ocks of birds such as cockatoos and udgerigars arrive and there are fish in ormally dry rivers. Kangaroos and emus wander over the plains. Large lakes such as ake Eyre are almost dry most of the time, ut after heavy rain they become vast heets of shimmering water.

Coastal cities and inland farms

'ighty per cent of Australians live in cities nd towns along the coasts. A smaller umber of people live in country towns or n scattered farms. The farms produce nainly beef, mutton and wool from sheep, nd wheat. These products are exported all ver the world. In the outback, where the oil is too poor to graze many animals in a mall area, the farms are enormous. Some f these 'stations', as they are called, are as arge as an English county.

n Tasmania, where the climate is cool and vet, there are many mountains, forests and akes, and not many people.

Papua New Guinea

Until 12,000 years ago, Papua New Guinea, Tasmania and Australia were all oined together. Then the sea rose and Papua and Tasmania became islands. Papua has a tropical climate with heavy rainfall and dense rainforests. There are nany high mountains and large rivers. The orests are full of wildlife, including the spectacular birds of paradise. The people ive in scattered villages, tending gardens in he forest and keeping pigs.

◄ Geysers erupt into the sunshine in Rotorua, New Zealand. The evaporating water leaves behind grey and yellow deposits of sulphur minerals.

A picture of the past

New Zealand became separated from Australia and the other continents a long time ago. It still has forests very much like those that covered large areas of the southern hemisphere 65 million years ago. In the warmer North Island there are huge kauri trees over 200 years old, strange tree ferns and flightless birds. Today, much of the forest has been cleared to grow vegetables, oranges, lemons and grapefruits, and apples, which are exported all over the world.

Both islands are in an earthquake zone. North Island has many active volcanoes. Around the town of Rotorua are numerous boiling mud pools, spouting geysers, bubbling hot springs, and mineral pools to bathe in.

South Island has a spectacular mountain range, the Southern Alps, with twenty peaks over 3,000 m (10,000 ft) and huge glaciers that have carved out deep fjords at the coast. Sheep are reared on the lower hills and grassy plains. New Zealanders are great lovers of the outdoors. Sailing, fishing and tramping in bush-clad hills are very popular pastimes.

◄ A small village in Papua New Guinea

The native people of New Zealand are called Maoris.

Eucalypts have leathery leaves to cope with the drought. Their leaves give off aromatic oils.

New Zealand's capital, Wellington, is built on at least 50 volcanic cones, some of them still active.

Pacific Islands

There are three main groups of peoples in the Pacific Islands, Melanesians in the southwest, Polynesians in the eastern Pacific, and Micronesians in Guam, Nauru and Kiribati.

Some of the beaches in Oceania are black. They are made of black volcanic sand.

There are about 10,000 islands in the central and southern Pacific Ocean, spread out over a huge area. They are often given the name Oceania. Most Pacific islands are coral reefs and the peaks of partly-submerged volcanoes. The climate is warm and moist, and the larger islands are mostly covered in tropical forest, except where this has been cleared for agriculture. The coasts have fine sandy beaches fringed with palm trees, which attract tourists to the larger islands.

Living on a hot spot

The Hawaiian Islands are the peaks of a chain of old volcanoes that stretch for 2,500 km (1,500 miles) across the northern Pacific. The largest island, Hawaii still has active volcanoes. The world's largest active volcano, Mauna Loa, is found on Hawaii. Its lava flows occupy more than 5,125 square kilometres (1,908 sq miles) of the island. The Hawaiian Islands have been formed over the last 25,000 years, one after the other from northwest to southeast. Each volcano is formed from molten rock produced above a hot spot in the Earth's crust under the Pacific. The floor of the Pacific Ocean is moving very slowly northwest. As one volcano moves away from the hot spot, a new one begins to form there. Today Kilauea is the main active volcano, and a new volcano is growing southeast of Hawaii. It is already nearly 5,000 m (16,000 ft) above the sea bed.

▲ This ring-shaped island is Kayangel Atoll. The atoll is a circular coral reef that has grown on the top of a submerged mountain. Fish and shellfish flourish in the warm waters of the central lagoon.

▶ The coast of Kauai Island, Hawaii. Volcanic slopes rise steeply out of the Pacific Ocean. Already the slopes have been eroded into gullies by the heavy rain sweeping in from the ocean.

Island life

About 1·5 million people live in Oceania. Between them they speak hundreds of different languages. Most people live by farming or fishing. The coconut palm provides food, drink and fibre. Coconut flesh is dried to make copra, which is pressed into coconut oil in factories. Coconut products are valuable exports. Rice, dal, yams, tapioca, fruit and fish make up a varied diet. Some islanders also rear pigs and poultry. Pineapples, bananas and other tropical fruits are grown for export. Nauru exports phosphate, which is used to make fertilizers.

Antarctica

◄ Pack ice breaking up in spring along the coast of the Antarctic Peninsula. The snowfields of the interior are visible in the background.

ntarctica is an almost circular continent ing over the South Pole. Most of the land covered by a huge ice sheet up to 3 km ick, which contains 90 per cent of all the orld's ice. Great mountain ranges lie eneath the ice. East Antarctica is a high ateau, but the west is a cluster of ountainous islands bound together by ice. few spectacular 'dry valleys' near cMurdo Sound are bare of ice. Floating e shelves fill many of the bays. Glaciers nd ice sheets break up at the coast to roduce icebergs.

Home of the blizzard

his is the world's coldest continent. The hite snow and ice reflect the Sun's heat ack into space. Temperatures range from 38°C to -25°C (-127°F to -31°F) on the ice ap, and -60°C to 15°C (-76°F to 59°F) ear the sea. Icy winds blow at up to 180 m/h (110 mph) down the slopes of the entral ice sheet, sweeping up loose snow to mock 'blizzards'. There is very little owfall. The ice sheets have built up over illions of years.

Very little life can survive here. Lichens, mosses and a few hardy plants grow near the coast. In summer, seals, seabirds and penguins thrive on the plentiful fish in the Southern Ocean. Whaling used to be an important industry here, but is now banned.

A continent for research

In 1959 twelve countries, including the USA, the former USSR and the United Kingdom, signed the Antarctic Treaty. They agreed not to make rival claims to territory, but to use the Antarctic only for peaceful research. Now 32 countries have accepted the treaty. Their flags stand in a ring around the South Pole

Scientists are the only permanent residents, and most of them stay only for the summer. Tourism is increasing, but poses a pollution problem. There are considerable mineral deposits in Antarctica, a large coalfield and probably petroleum, too, but these are too expensive to exploit at present.

The world's coldest temperature, -89·2°C (-128°F), was recorded at Vostok, in Antarctica.

150 million years ago, Antarctica had a warm climate and lush forests.

The emperor penguins spend the whole winter in Antarctica, sitting on their eggs.

Antarctica is the highest continent in the world, averaging 2,100 –2,400 m (7,000–8,000 ft) above sea level. But without its ice sheet it would be only 460 m (1,500 ft) high, and much smaller.

Waste disposal is a problem in Antarctica. Nothing rots at these low temperatures. Visiting scientists have to take most of their waste home with them when they leave.

Northern Africa

Northern Africa is a vast region of tropical and subtropical desert, high mountains and large river basins. The Sahara Desert occupies most of the region. It forms a series of flat rocky plateaus, occasionally broken by jagged mountains of harder rock, such as the Hoggar and Tibesti ranges. In the northeast, the greener Atlas Mountains capture rain from winds blowing in from the Atlantic Ocean. To the east are the valleys of the Nile and its tributaries. In the south is the inland basin of Lake Chad, fed by rivers from the surrounding highlands.

▼ The inhospitable Hoggar Mountains rise above their desert surroundings in near Tamanrasset in Algeria. Much of the North African desert is rocky.

Oil and natural gas reserves occur in the north of the region. Libya has half of Africa's reserves of oil. Tunisia also exports oil and phosphates, and Niger sells uranium. But most of the mineral reserves of the region have not yet been exploited.

The Atlas Mountains

The Atlas Mountains are snow-capped peaks rising above 4,000 m (13,500 ft). They extend for over 2,000 km (1,200 miles) through Morocco, Algeria and Tunisia. The rain tends to fall in sudden downpours, cutting deep gorges through the mountains. Where these gorges open out onto the lowlands, small conical hills of sediment have been deposited as the fast-flowing rivers spread out over the plains. On the higher slopes are the famous Atlas cedars, while in the wetter areas are forests of cork oaks, a profitable export.

Many people live in the mountain villages, some of which look more like forts than villages. Cattle, sheep and goats graze in the mountains in summer, and are led back to the valleys in winter. The green valleys have fruit orchards and groves of figs and olives. With irrigation, cereals are grown in summer, and vegetables in winter. Many people work in the leather industry. The mountains are becoming popular for tourism and winter sports.

The Sahara Desert

The Sahara is the largest desert in the world. There are high dunes in places, but most of the desert is rocky. Wind-blown sand has blasted the mountains into fantastic shapes. Some places go for years without rain. The skies are always clear, and summer temperatures are among the highest in the world, up to 43°C (109°F). Winters are cooler, and nights can be cold and frosty.

Only about 2 million people live in the Sahara, mostly around the oases. Here they irrigate small gardens and grow date palms, millet, barley and wheat, fruit and vegetables. Outside the oases, nomads roam the desert with their goats, sheep and camels.

Thousands of years ago the Sahara was much wetter. Valleys that are dry today contained rivers, and people lived there. They shared the land with elephants, giraffes, buffalo and hippos. These people left behind pottery and tombs. Stone Age fish-hooks have been found on dried-up lake bottoms, and there are the remains of large trees buried in the sand.

◄ The Tifnoure Valley in the High Atlas Mountains of Morocco. The low stone walls built across the valley floor probably help to trap water in the little fields when the melting snow swells the river in spring. In the background, cereals are being grown on hill terraces. The green valley contrasts with the arid mountain slopes.

The Sahara desert suffers from violent dust storms. The winds that cause these storms are called simoon, sirocco or khamsin.

The name 'Sahara' is Arabic for 'desert'.

When the wind blows over the high Sahara sand dunes the sand 'smokes' and the wind makes a 'drumming' sound. Local people call it the Spirit of Raoul, the drummer of death.

The highest temperature recorded in the Sahara was 84°C (151°F). Yet in the Tibesti Mountains the temperature can fall to -15°C (5°F) in winter.

► A herd of goats is being taken in search of pasture in a Sahel landscape already stripped of most of its trees. Overgrazing of the thin pasture in this semi-desert region soon destroys the vegetation. This results in soil erosion and the advance of the desert.

From the desert to the rainforest

In the south, the Sahara merges into a vast semi-desert grassland called the Sahel. In good years, the grasses support cattle, sheep and goats. But there is a constant battle between desert and grassland. In drought years, overgrazing exposes the sandy soil to the wind, and the desert advances. As rainfall increases, true savannah takes over with scattered acacia trees and, in the far south, strange baobab trees that store water in their swollen trunks.

Watered by the Nile

The land beside the Nile stands out as a green ribbon in the desert. In the rainy season, the river overflows its banks, spreading fertile silt across the land. In Sudan the water feeds the great papyrus swamps of the Sudd, where local people graze their cattle on the lush vegetation.

A lot of people live on the floodplain and the delta. The land is intensively farmed, fed by a network of irrigation canals. Oxen, donkeys and camels are used for turning wheels to pump water into the fields. Maize and vegetables are grown, as well as dates, olives, water melons and cotton for export. The river also provides good fishing.

The Nile is an important highway, carrying passengers and cargoes to and from Cairo. The many remains of ancient Egypt,

▼ Irrigating fields on the west bank of the Nile. The water-wheel which does the pumping is turned by oxen, but nowadays much irrigation is done with electric pumps.

A vast inland lake

Lake Chad is all that remains of a much larger ancient sea. Crocodiles and hippos live here. There are plenty of fish, especially tilapias, which are good to eat. Around the lake shores, people build dykes and dams to reclaim small fields from the swamps. Here they grow rice, cereals, beans, okra and water melons. South of the lake the climate is wetter, and farmers can grow crops like cotton and peanuts for export. Cattle are reared on the dry grassy plains.

The cultivated coasts

Many North Africans live on the coastal plains. Here they grow grapes for wine, and plant orchards of apricots and almonds, and groves of citrus trees, olives, figs and date palms, as well as vegetables, rice and cereals. They also produce cash crops such as tobacco, cotton, sunflower seeds and sugar cane.

The Egyptian state is one of the oldest in the world. It dates back to 3100 BC.

Ethiopia once had a powerful empire called Abyssinia.

The swamps and marshes along the Nile valley contain vast beds of a reed called papyrus. The ancient Egyptians invented a way of making a kind of paper, also called papyrus, from these reeds.

▼ **A landscape near Adi Shah in southern Tigray, Ethiopia. Long, narrow hill terraces have been ploughed ready for planting after the next rains. Droughts are common in this part of the world, and have led to bad famines in the past.**

including the Pyramids, support a profitable tourist industry. Egypt also makes money by charging ships to use the Suez Canal, so they can avoid the long and dangerous passage around the tip of Africa.

The Horn of Africa

Ethiopia, Djibouti and Somalia are often called the Horn of Africa, because of their shape on the map. This is an area of high, cool uplands and hot, dry semi-desert lowlands. In the uplands people farm the land, and produce coffee for export. On the lowlands they rear livestock where there is enough pasture.

Droughts in the 1970s and 1980s ruined crops and cattle died. Many people starved. Djibouti and Mogadishu are important ports, trading with Arabia and India.

The African Rift Valley, a great split in the Earth's crust, runs through the Ethiopian Highlands. Rainfall in the Highlands feeds the waters of the River Nile. The Red Sea also lies in the Rift Valley. It has some of the world's most beautiful coral reefs.

Central Africa

▲ Zebra and gazelle grazing on the African savannah in Amboseli National Park, Kenya. The tall acacia trees provide the only shade in these grasslands. In the background is Mount Kilimanjaro, an old volcano 5,895 m (19,340 ft) high.

Central Africa has a great variety of landscapes, from the dense rainforests of the Zaïre basin to the semi-desert grasslands of the Sahel, and the deep lakes and active volcanoes of the African Rift Valley. Several large rivers drain into the Atlantic Ocean. The Senegal, Gambia, Volta, Niger and Zaïre (Congo) rivers all have large river basins which are very fertile and densely settled.

In between the river basins are high plateaus and mountains, such as the Guinea highlands and the Cameroon highlands. To the east the land rises toward the East African plateau. On the borders of Zaïre and Uganda the Ruwenzori mountains, often called the Mountains of the Moon, are over 3,000 m (10,000 ft) above sea level.

Snow-capped peaks and fiery volcanoes

This is the edge of the African Rift Valley, a huge branching split in the Earth's crust which is still growing today. As the continent tears apart, molten lava wells up along the edges of the rift. It forms volcanoes like the Virunga Mountains in Rwanda and Zaïre, which are still active today. The highest mountains in Africa, Mt Kenya and Mt Kilimanjaro, are extinct volcanoes well over 5,000 m (16,000 ft) high. The tallest peaks are snow-capped, with many small glaciers, even though the mountains are close to the Equator.

In the Rift Valley itself are a number of

arge lakes, such as Lakes Rudolf, Magadi,
Jyala, and Tanganyika. Most of them are
ery deep. Lake Victoria, the source of the
Liver Nile, lies on a plateau and is quite
hallow. The eastern branch of the Rift
alley is very dry and hot. Many of its
akes, such as Lake Magadi, are very salty
ecause of the high evaporation rate.

Life in the Rift Valley

The mountains have a cool climate with
igh rainfall, but the troughs in between
re very hot and dry. There are lush forests
round Lake Victoria, but semi-desert in the
leeper valleys. Fishing is the main source of
ncome for villages on the shores of the
reshwater lakes.

The volcanic soils are very fertile, and there
re many villages on the mountain slopes
elow about 2,100 m (7,000 ft). Millet,
naize and bananas are the main food
rops. This is one of Africa's chief coffee-
rowing areas; cotton, sisal, tea and cashew
uts are also exported. Cattle are raised on
he dried grasslands.

Many tourists come to see the dramatic
cenery and the wild animals that roam the
avannahs. The elephants, giraffes,
ntelopes, zebra, wildebeest and lions are
protected in many national parks, such as
he Serengeti in Tanzania and the Masai
Mara in Kenya. The Virunga and Volcanoes
National Parks in Zaïre and Rwanda are
nome to mountain gorillas.

Tropical Africa

The rest of Central Africa is lower land.
Close to the Equator it is hot and wet all
year round. Dense tropical rainforests grow
where there is more than 125 mm (50 in)
rainfall a year. Many rare animals, such as
gorillas, okapis, forest elephants and
leopards still survive there. The local people
are mostly farmers, clearing the forest to
produce just enough food for their own use.
A few countries export timber, but much of
the forest is a long way from the coast and
transport is difficult.

The upper slopes of the high mountains of East Africa have some very strange giant plants. Here, heathers, groundsels and lobelias grow to tree size.

Along the coast of Benin, whole villages are built out into the sea on stilts.

The jungles of the Zaïre basin are so vast that parts of them have still not been explored.

Nigeria, Gabon and Congo export
petroleum. Senegal is the centre of the
African film industry, and also exports large
amounts of phosphate for fertilizers. Sierra
Leone and the Central African Republic
produce diamonds, while Liberia is a major
exporter of iron ore. Guinea exports bauxite
(aluminium ore), and Zaïre is a leading
copper exporter.

▲ Fishermen bring their catch ashore at Elmina, in Ghana. Fish is an important source of protein in many African countries.

◄ Lake Bogoria, in Kenya, lies in the African Rift Valley. Huge flocks of flamingoes feed in the shallow water near the shore.

▲ **The River Ndian flows through dense rainforest in the Korup National Park, Cameroon.**

The Zaïre (Congo) river is the second longest in Africa. The river is 4,670 km (2,900 miles) long, and its basin covers 3,457,000 sq km (1,335,000 sq miles).

Lake Tanganyika is the second deepest lake in the world. Its maximum depth is 1,435 m (4,710 ft), over 600 m (2,000 ft) below sea level.

▶ **Women collect water in a village in Rwanda. Much of the forest around this village has been cleared to plant crops.**

Lush river basins

A lot of people live in the fertile river basins and along the coastal plains. After the rains the rivers flood, and the water is trapped behind dykes or led into small canals. The main food crops are cassava, yams and the cereals sorghum, millet, maize and rice. Coffee, cocoa, palms (for palm oil), groundnuts (peanuts), cotton and, in Nigeria rubber, are grown for export. Many

people are fishermen, and surplus fish are exported. Liberia has the world's largest merchant navy. The big rivers form major highways for transporting goods. This has allowed economic development far inland.

The savannahs

North and south of the Equator rainfall decreases and the weather becomes more seasonal, with definite wet and dry seasons. The jungle gives way to woodland savannah, a mixture of grassland and woodlands of deciduous trees (trees that shed their leaves during the dry season). Where there is even less rainfall, the savannah becomes more thorny and scrub-like. Spiny acacia trees are dotted over the grasslands, and there are huge baobabs up to 30 m (100 ft) tall, with swollen trunks that store water.

The savannahs are dotted with small villages, where farmers just manage to make a living from the land. Where there is enough rainfall, people keep cattle. In the drier areas, cattle herders like the Masai tribe of Tanzania live as nomads.

Large herds of zebras, wildebeest and gazelles also roam the savannahs. Giraffes browse the thorny acacia trees. Lions, leopards, cheetahs, hunting dogs and hyenas prey on the migrating herds, and vultures soar overhead, looking for carcasses. The grass seeds support huge flocks of small seed-eating birds.

Southern Africa

outhern Africa is a region of upland
plateaus, the worn-down remains of
ancient mountains. River gorges divide the
plateaus into large blocks and flat-topped
mountains. The Zambezi, Limpopo and
Orange rivers have carved out wide valleys.
In the southeast are the jagged mountains
of the Drakensberg Plateau ('the Dragon
Mountains'). The high plateaus shield the
great deserts of the Kalahari and the Namib
from rain-bearing winds. Madagascar, too,
is a hilly country. It became separated from
the continent millions of years ago, and has
many unusual rare animals and plants.

The region has a seasonal climate. The
north is almost tropical, with rain falling
mostly in summer. The south is cooler and
has less rain. Here, winter is the wettest
season.

The veld

Most of Southern Africa's plateaus and
plains are covered in grassland known as
the 'veld'. This is one of the world's oldest
inhabited regions. The land above 1,200 m
(4,000 ft) is called the 'highveld'. In
Lesotho it rises above 3,000 m (10,000 ft)
in places, and frost is common on winter
nights. The highveld soil is thin and poor,
and it is easily blown away. There are
many scattered pans, large flat salty areas
that contain shallow lakes in the rainy
season. The 'lowveld', below 120 m (500 ft),
is wetter and more fertile. In Zimbabwe, the
veld is more wooded, especially along the
river valleys.

Cattle, sheep and goats are raised on the
veld. South Africa produces half of Africa's
wool, mostly from merino sheep. In
Lesotho, angora goats are reared for wool
and mohair. Maize is the main food crop.
Sugar cane, citrus fruits, tobacco and tea
are grown near the moist east coast. Cotton
is an important cash crop, too, especially
on irrigated land near the rivers.

The Zambezi river

The Zambezi valley is an important trading
route. There are many fishing villages
along its banks. The river has been
dammed to create two great lakes. Lake
Kariba is shared by Zambia and Zimbabwe,
and Cabora Bassa is in Mozambique. The
dams generate hydroelectric power, and the
lakes provide water for irrigation. The
Zambezi river, Victoria Falls and Lake
Malawi attract many tourists, and so does
the plentiful big game to be seen in the
region's many national parks.

▼ The Victoria Falls, on
the border between
Zambia and Zimbabwe,
is one of Africa's
greatest tourist
attractions.

▲ Arid grassland in Zimbabwe. In the drier parts of the veld, many plants are adapted for saving water. The strange baobab tree stores water in its huge swollen trunk, and the cactus-like euphorbias also have succulent water-storing stems.

Madagascar is the world's fourth largest island.

The Bushmen or San people were probably the earliest inhabitants of southern Africa. They live in the Kalahari in wandering groups of 25 to 30 people, gathering wild berries and roots and hunting game. Today, more and more of them are working on farms, and giving up their traditional way of life.

▶ The town of Windhoek, capital of Namibia, is a mixture of old colonial buildings and modern sky-scrapers, complete with satellite dishes.

The great deserts

The Kalahari desert occupies much of the hot, dry heart of the region. It is a huge basin over 900 m (3,000 ft) above sea level, mostly covered with gold or red sand. In places there are dry river valleys, formed thousands of year ago when the climate was wetter.

In the north is a great swamp, the Okavango, where the Okavango river ends in a large inland delta. This is a haven for wildlife such as lions, antelopes, elephants, giraffes and zebra. In the northeast is the vast Makgadikgadi Pan, where several rivers evaporate and disappear into the desert.

Much of the area is covered in tussocky grassland with scattered trees and thorny shrubs. Most of the local people are cattle farmers. They rely on wells and boreholes for water. Cattle are a sign of wealth and status, and people tend to keep so many that overgrazing destroys the grasslands. In fact, goats provide most of the meat and milk. Maize and sorghum are grown on a small scale, but wild plants and animals form an important part of the diet.

Coasts dry and wet

Even fewer people live in the Namib Desert, which fringes the west coast. The main source of water here is the fog that rolls in from the sea. This is called the Skeleton Coast, because so many ships have been wrecked here and lie buried beneath the dunes. The moist east coast has mangrove swamps and fertile coastal plains. The chief food crop here is cassava, which is used to make tapioca. Cashew nuts, tea and cotton are grown for export. Coconuts are an important crop in the Comoros islands, and in Mozambique and Tanzania.

The 'gold mine' of Africa

Southern Africa is a major source of many valuable minerals. Some of the world's largest deposits of diamonds are found in Southern Africa, especially in South Africa and Namibia. Madagascar has gemstones such as garnets, tourmaline and zirconium. South Africa also has major deposits of platinum, uranium and antimony and, together with Zimbabwe, is a leading exporter of gold and chromium. In Zimbabwe and South Africa, iron ore supports important steel and engineering industries. Zambia lies in the Central African Copper Belt.

Alaska, Canada and Greenland

◀ Inuit at the edge of the ice in Greenland. The Inuit are the main native people of the region. They used to be nomads, hunting fish and seals with harpoons from their kayaks (canoes), and living in shelters made of ice. They travelled over the frozen land on sledges pulled by teams of husky dogs. Today most Inuit live in towns and work in fish factories.

Canada is one of the world's largest countries. It stretches from the frozen Arctic Ocean in the north to the prairies in the south. The west coast is rocky, with spectacular glaciers and ice-carved scenery. Behind the coast are the jagged peaks of the Coast Mountains and the Rocky Mountains which are still rising. There are many active volcanoes. Labrador and nearby Greenland are also mountainous and rocky. The centre of the region is flatter, broken up by a series of large lakes and thousands of smaller ones.

The Arctic Ocean

Most of the Arctic is a huge frozen ocean. In winter it is dark for most of the time, and very cold. In summer the Sun shines for 24 hours: the Arctic is the land of the midnight Sun. For a few months the edge of the Arctic Ocean melts. The Arctic Ocean is rich in fish, shrimps and other shellfish, which are food for seals, walruses, whales and seabirds.

Oil and natural gas have been found near its shores, but extracting it poses a threat to this fragile environment. Oil spills could spread for great distances under the ice, where they would be almost impossible to remove. Oil blocks out air from the water below, killing the marine life. In these low temperatures, oil can take many years to break down.

▼ The Yukon Territory, in north-west Canada, is a vast area of mountain ranges, plateaus and forests. It contains Canada's highest peak, Mount Logan (5,951m).

The frozen north

Much of the Arctic Ocean is frozen all year, and most of Greenland is covered in a thick ice cap around 1,500 m (5,000 ft) thick. In summer, jagged icebergs break away from the melting ice cap and glaciers. Frozen, treeless tundra fringes the Arctic Ocean. Only a thin layer of soil is unfrozen in summer. The tundra is covered in lichens, mosses and low-growing plants and shrubs. These are food for caribou (reindeer) in summer. Millions of insects breed in the melt-water pools. They attract huge numbers of birds that migrate to the tundra to breed. Arctic foxes prowl the tundra in search of eggs and young birds.

In autumn the birds leave for warmer climates, and the caribou migrate south to the shelter of the coniferous forests. Here, the snow is softer, and the caribou can scrape it away to find grazing below. The polar bears move out onto the newly-formed sea ice to hunt seals.

The dark forests

South of the tundra is a vast lake-studded belt of coniferous forest. Further south, where summers are longer and winters milder, the forest is a mixture of evergreen trees like hemlocks, and deciduous species such as maples, elms and oaks. With the first freezing weather in October, the maple leaf, Canada's symbol, turns red and the forest appears to be on fire. Canada and Alaska have important timber industries, producing paper and plywood for export. The logs are floated down the rivers to the sawmills.

The prairies

The flat grasslands of the prairies form the main wheat-growing area of North America. The huge fields stretch to the horizon. Wheat is exported all over the world. Farming is on a very large scale, using heavy machinery and not many workers.

Winters here are cold, and snow covers the land from November until April. Snow ploughs push snow off the city streets and salt melts ice on the pavements. Winters can be difficult in the cities, but fun in the country. People ski and skate and travel over frozen lakes in snowmobiles. Summers are warm enough for gardening and swimming in the lakes.

Most people live in the cities. Along the Atlantic and Pacific coast cities have grown up around safe harbours. Vancouver, on the Pacific Coast, grew up as a port at the western end of Canada's first continental railway. Cities are strung out along the railways, main roads and the Great Lakes – St Lawrence Seaway.

The great lakes

The St Lawrence River and the St Lawrence Seaway link the Great Lakes with the Atlantic Ocean. Half Canada's population and most of its industries and large cities are found along this water system, which is a major highway for exports and imports. Gold, nickel, asbestos and platinum from Canada's mines are important exports. Fishing is important around the Great Lakes, and off the coasts of Newfoundland and Greenland.

It is possible to travel right under the North Pole in a submarine. The first ship to make the crossing was the nuclear-powered submarine the *USS Nautilus*. In 1959 another American submarine, the *USS Skate* broke through the ice and surfaced at the North Pole.

At the North Pole the Arctic Ocean is 4,087 m (13,419 ft) deep.

Polar bears have been known to travel overland in winter from Greenland to the islands of Svalbard off the north coast of Norway.

▼ Huge grain silos line the shores of Thunder Bay in Ontario, Canada. Grain is a major export. The railway wagons bring in grain from many parts of the country. The grain is stored in the silos before being loaded onto ships for the journey through the Great Lakes and the St Lawrence Seaway to the Atlantic Ocean.

United States of America

◄ **Winter in the Grand Teton National Park in Wyoming. The rocks have been carved by rain and frost into deep canyons and rocky pinnacles.**

The Rocky Mountains are about 4,800 km (3,000 miles) long, and several hundred miles wide. The Rockies contain many gorges and canyons. The Grand Canyon of the Colorado River is, in places, about 1·6 km (1 mile) deep. Nearly 3 million people visit it every year.

The United States is a country of great contrasts. There are hot deserts and snow-capped mountains, dense forests and vast prairies, sandy beaches and alligator-infested swamps, spectacular wilderness areas and big industrial cities.

Mountain country

One third of the country is occupied by the great mountain ranges that run from north to south down the western side of the continent. The greatest of these are the Rocky Mountains, which contain some of the most dramatic scenery in the country.

Some mountains have gentle slopes and rounded tops, but others are steep with jagged rocky peaks, many over 4,000 m (13,000 ft) above sea level. Among the mountains are wide valleys, plateaus, lakes and rivers.

The Rockies began to form 190 million years ago, and are still rising slowly today. Many great rivers start in the Rockies, and some have cut deep valleys and canyons. The winds blowing from the west bring rain to the western slopes, leaving the central and eastern Rockies and the Great Plains dry.

Earthquakes and volcanoes

Along the west coast, two of the Earth's great crustal plates are in collision. The Pacific Ocean floor is being forced under the North American continent. This causes earthquakes, especially in California. A major fracture line runs all the way from San Francisco to the Gulf of California. This is the San Andreas Fault. As the rocks push past each other along this line they set off earthquakes.

To the west of the Rockies are more mountains, the Sierra Nevada and the Coast Ranges. Some are the remains of a chain of volcanoes, some of them still active. Mount St Helens exploded unexpectedly in 1980. It blew the top 400 m (1,300 ft) off the mountain. Sixty-five people were killed and forests were destroyed by the blast up to 20 km (21 miles) away.

Life in the Rockies

▲ Ruled by the car - the city of Los Angeles lies between the Pacific Ocean and the San Gabriel Mountains.

▶ Giant cacti, prickly pears and blooming ocotillo in the desert at Saguaro National Monument, Arizona.

One of the worst earthquakes caused by movement along the San Andreas Fault happened in 1906 in San Francisco. The earthquake and fires that were started by it destroyed much of the city and killed 70 people.

The Rocky Mountains contain deposits of metals such as iron, silver, gold, lead and zinc, as well as uranium, phosphates and other salts. There is also coal, oil and natural gas. Some of the rivers have been dammed for hydroelectric power and irrigation.

The higher peaks are capped with snow and glaciers. The alpine meadows are full of wild flowers. Below the meadows, forests cover much of the mountain slopes. Many large mammals live in the mountain forests: bears, cougars (mountain lions), elk (moose), deer, and wild mountain goats and sheep. An important source of income is tourism, including skiing, fishing, and walking in the many national parks.

Plateaus and basins

Many people live in the great Central Valley of California, between the Sierra Nevada and the Coast Ranges. The climate is warm and wet, ideal for growing citrus fruits and grape vines. Californian wines are world-famous. Oil and natural gas are found along this coast and offshore.

Between the Sierra Nevada and the Rockies is a huge area of basins and plateaus. This is an arid region, cut off from the rain-bearing westerly winds. The high plateaus are very cold in winter, but in the south there are hot deserts like the Arizona and Sonoran Deserts. Death Valley in California is the hottest and the lowest place on the continent. Part of the valley is 82 m (282 ft) below sea level. Temperatures here can reach 57°C (134°F) in the shade, and 88°C (190°F) on the ground.

Flatlands

Large parts of central USA are vast flat plains. Two hundred years ago thousands of buffalo grazed these grasslands. Today there are few buffalo left. The grasslands have been ploughed up and planted with crops, especially wheat. The USA is the world's largest exporter of wheat. Summers are hot and dry, but winters can be bitterly cold as icy winds sweep down from the Arctic. The drier south, particularly Texas, is 'cowboy country', where cattle are reared on huge ranches.

Several large rivers meander across the plains. The Mississippi-Missouri river system is the fourth longest in the world. Many dams have been built to control the seasonal floods. The rivers provide water for cities and for crops, and are used for transport.

Canal networks link industrial centres. The New York State Barge Canal connects the Hudson River to Lake Erie. The Illinois Waterway links Chicago and Mexico, and the Gulf Intra-coastal Waterway links Mexico and Florida.

The coasts of Florida and the Gulf of Mexico sometimes suffer from hurricanes between June and October.

The west coast is pounded by the large breakers of the Pacific Ocean. Its beaches are excellent for surfing.

▲ The wheat harvest on the prairies in Kansas. The large fields allow heavy machinery such as combine harvesters to be used.

The swampy south

The southeastern USA, around the Gulf of Mexico, has a humid and subtropical climate. The southern part of the Florida peninsula is a wilderness of wetland, forest and mangroves known as the Everglades. Alligators, ibises and pelicans live in these swamps. Much of the Gulf coast is swampy. The warm climate and plentiful water supply is well suited to growing cotton, rice, sugar cane and tobacco. Oil and natural gas are found around the Gulf.

The industrial north-east

In the eastern United States are the Appalachian mountains. These are quite different from the Rockies: they are older, lower and less jagged. Their slopes are heavily forested. Many industries have developed around local iron and coal deposits, but oil, natural gas and hydroelectric power are the main energy sources used today. Iron, coal and engineering goods such as machinery, aircraft and motor vehicles are important exports.

The farms of the north-east are much smaller than those on the western plains. They are often mixed, growing cereals and vegetables, and rearing cattle, pigs, sheep and poultry. Timber and timber products are a major source of income for the USA. The country also has one of the world's largest fishing fleets.

Contrasting coasts

The east and west coasts of the United States are very different. The west coast is pounded by the large breakers of the Pacific Ocean. Its beaches are excellent for surfing, and its rugged headlands make for some picturesque scenery. Off the California coast underwater forests of kelp (a large brown seaweed) are breeding grounds for many kinds of commercial fish and shellfish, such as abalones. They are also home to the sea otter. The kelps themselves are harvested and used to produce a variety of chemicals and fertilizers.

Along much of the east coast there are sand dunes, spits and offshore islands which shelter calm lagoons. Saltmarshes line muddy estuaries, which support large numbers of wading birds. Along the warm coasts of Florida and the Gulf of Mexico coral reefs and mangrove swamps teem with life, and the white sand beaches attract large numbers of tourists from all over the world. These coasts sometimes suffer from hurricanes between June and October.

The forests of the west coast contain some of the tallest trees in the world, the Californian redwoods. These can grow to well over 100 m (330 ft), and some are over 3,000 years old.

Along the coast of the wet northwest are rainforests. These are temperate rainforests. Like tropical rainforests, they are damp, lush evergreen forests thick with trailing creepers, mosses and ferns.

▶ A steel plant in Pennsylvania. Much of the United States' manufacturing industry is based in the north-east, close to mineral and fuel resources. Here, there is a good network of roads and railways for transporting raw material and finished goods.

Central America and the Caribbean

◀ Union Island, in the Grenadines, is one of about 600 small islands between Grenada and St Vincent in the eastern Caribbean Sea. The pale turquoise areas are coral reefs.

Central America is a wedge-shaped neck of land that joins North America to South America. It has a backbone of mountains and volcanoes, rising to over 5,000 m (16,000 ft) above sea level. Some of the volcanoes are still active, and earthquakes are common. A chain of islands over 3,000 km (1,900 miles) long separates the Caribbean Sea from the Atlantic Ocean. These islands are sometimes called the West Indies, because when Christopher Columbus found them in 1492 he thought he had reached the Indies in Asia.

Most of the region has a warm tropical climate with plenty of sunshine. Most rain falls in summer and autumn, and winters can be very dry. Hurricanes sometimes whirl in from the Atlantic Ocean during the rainy season, causing havoc along the coast. Many tropical crops grow well here. Sugar cane, bananas, coffee, cocoa, spices and tobacco are all grown for export, often

on land that was once covered in tropical forest. Cuba is famous for its Havana cigars. Some of the islands have valuable aluminium ore, and there are oilfields around the shores of the Gulf of Mexico and Trinidad.

Cactus country

Northern Mexico is an upland region of desert and semi-desert, warm by day, but cool by night. Cacti of all shapes and sizes dominate the landscape. Baja California, on the Pacific coast, has a rich sea life. Whales come here to give birth, and there are thousands of seals and seabirds.

From turtles to toucans

Much of mainland Central America is covered in rainforests, which contain a wealth of wildlife, including jaguars, ocelots, howler monkeys, toucans and

Belize has the world's second largest coral barrier reef.

The largest of the Caribbean islands, Cuba, is over 1,100 km (700 miles) long and has a population of 10 million.

Mexico City is the largest city in the world, with over 18 million people (including those who live in the suburbs). It is also one of the world's most polluted cities.

quetzals. Tourism is a growing industry. The wildlife, the fascinating ruins of the ancient civilizations of the Mayas and Aztecs, and the glorious beaches of white coral sand are the main attractions. Sea turtles also spend time here, laying their eggs in the warm sand, which acts as an incubator.

The volcanic soil is very fertile, and the lowlands are ideal for plantations of sugar cane, bananas and other tropical crops. Some of the forest has been burned to provide grazing for cattle. As in South America, most of the land is owned by a very few wealthy people, and the people who actually work on the land are very poor.

Island life

The Caribbean islands extend from the coast of Mexico past the southern tip of the United States to Venezuela. Some are just small coral reefs, called cays, that scarcely rise above the water. The larger islands are really the tops of partly submerged mountains, some of them old volcanoes. The larger islands are covered in tropical forest, which has been cleared for farming in the lowlands. Most of the islands are fringed by coral reefs.

The Panama Canal

The Panama Canal was built across the narrow isthmus of Panama to allow ships to travel from the Atlantic Ocean and the Caribbean Sea to the Pacific Ocean without having to sail around the stormy tip of South America. The canal is 82 km (51 miles) long, and has six enormous locks which carry the water up to 22 m (72 ft) above sea level. It was completed in 1914, but at a great human cost. Over 25,000 workers died from disease or accident while the canal was being built. Today the canal is silting up with soil washed down from the surrounding hills, which have been cleared of forest.

▲ Rainforest in the Udiabi Reserve, Panama. Central America still has large tracts of unspoilt rainforest. Many of the reserves now attract 'nature tourists'.

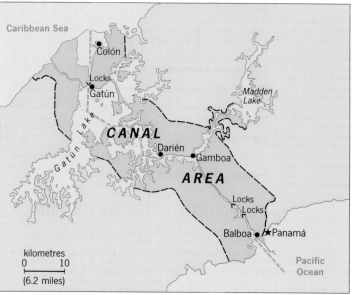

Tropical South America

The vast basin of the River Amazon forms the heart of tropical South America. Most of it is covered in tropical forest. To the west are the Andes, the world's longest mountain range. These young mountains are still rising today. Snow-capped volcanoes mingle with jagged peaks, and earthquakes are common.

The lowlands are hot and wet all year round, but where the land slopes gently up to the surrounding hills and plateaus, the rainfall decreases and the weather becomes more seasonal. The dense jungles give way to drier forest, then to savannah grasslands, called 'llanos' in the north and 'chacos' in the south.

Oil fields contribute to the economies of Venezuela, Ecuador and Peru. Brazil is a major industrial nation, with engineering industries based on steel. In other countries ores of metals such as aluminium and tin are exported.

Over 15,000 tributaries and subtributaries feed the Amazon river. More than 200 different kinds of fish live in the River Amazon.

Many seabirds live along the Pacific coast. The islands where they nest are thickly covered in bird droppings, called guano. This is collected and used to make fertilizer.

There is a huge gap between rich and poor in most of South America, with a few people owning most of the land and the money.

◄ The meeting of two tributaries of the River Amazon, the Mishagua and the Urubamba, in Peru. The River Mishagua is coloured brown by a heavy load of soil eroded from bare hill slopes.

▲ An Uru Indian punts his reed boat across Lake Titicaca, 3,811 m (12,500 ft) up in the Andes. These Indians build their homes on floating islands of reeds.

▶ Farmers planting potatoes on the Altiplano in Bolivia, a cold bleak plain between two mountain ranges in the Andes. This is the home of wild potatoes, from which all our domestic potatoes are descended.

The Amazonian rainforest

The Amazonian rainforest and the jungles of the Orinoco River basin are home to a huge variety of plants and animals, including jaguars, monkeys, sloths, toucans and macaws. Native Indians still live in the forest in small, scattered villages. They hunt wild animals, fish in the rivers, gather plants for food and medicines, and grow cassava and vegetables in small clearings in the forest.

Large areas of the Amazon rainforest are being felled by loggers, who export the valuable timber. The loggers' roads open up the forest to poor settlers from the cities, who are invading the forest in large numbers. They burn down the forest to plant crops. Cattle ranchers are destroying even larger areas and driving out the Indians. But the soil here is thin and poor. Without the tree cover it is soon washed away by the heavy rain. In any case, it cannot support crops or pasture for more than a few years.

Once the forest is cleared, it may take hundreds of years to grow back. Yet many valuable products, such as rubber, nuts, varnishes, lacquers, and many medicines and pharmaceuticals could be extracted from the forest without destroying it.

Coca, coffee and giant tortoises

The Andes are the second highest mountain range in the world. Between the mountain ranges are high cold plains. On the Altiplano in Bolivia is the world's highest city, La Paz. Lake Titicaca is the highest navigable lake in the world. The local Indians use reed boats to travel to and from their island homes on the lake.

Sheep, goats and alpacas are reared on the mountain grasslands. On the moist eastern slopes the main cash crop is coca, the basis of the illegal cocaine industry. In the foothills farmers grow coffee, sugar cane, cotton and bananas for export. The Pacific coast once had a large fishing industry based on anchovies. Then the climate changed and fish stocks crashed. Thousands of tourists visit the Galapagos Islands, a cluster of volcanic islands with many unique plants and animals, such as giant tortoises and marine iguanas.

Cattle country

To the north and south of Amazonia are large plateaus. In the north the rugged Guiana Highlands rise above 2500 m (8000 ft). Few people live here. By contrast, many people live on the Brazilian Plateau and the Mato Grosso. This is higher but less rugged country. Coffee is the major export crop. Other crops include sugar cane, tobacco, maize, bananas, rice and oranges. The grasslands are cattle country. Cattle are reared for export on very large ranches.

Temperate South America

Temperate South America extends from the subtropics almost to Antarctica. It is a region of mountains and grassy plains, temperate rainforests and some of the bleakest deserts in the world. Oil is found in several parts of the region, and there are substantial mineral deposits.

The climate is dramatically affected by the high Andes mountain range. In the north the Andes block the south-east trade winds. The Rio de la Plata basin gets plenty of rain, but the Atacama Desert on the east coast gets almost no rain at all. This is the world's driest desert, and also its cloudiest. Cold ocean currents cause cloud and fog to form all year round. Farther south, the mountains are in the way of the westerly winds. The western slopes of the Andes are covered in lush temperate rainforest, but east of the Andes is the cold Patagonian Desert. In the far south, the mountains are capped by snowfields and glaciers.

South America stretches much further south than any other continent. The tip of the continent is only about 1,000 km (600 miles) from Antarctica.

The Atacama Desert in Chile is the driest place on Earth. Rain has never been recorded in some places.

The pampas have their own special wildlife. Giant anteaters hunt for termites and bands of large ostrich-like birds called rheas roam the plains.

The ice fields and glaciers of southern Chile owe their existence not so much to cold winters as to cool summers which do not allow much snow-melt.

Chile is the longest country in the world, measured from north to south. It extends from the hot tropics to within 850 km (300 miles) of the Antarctic ice.

◀ **The Atacama Desert in Chile, with the snow-capped Andes beyond. Only coarse tussocky grass and low-growing 'cushion plants' can grow in this bleak, windswept landscape.**

Bleak mountains and lush forests

The Andes are high and rugged. There are several active volcanoes, some towering above 6,000 m (almost 20,000 ft). Ice and rivers have carved deep gorges and valleys. In the south the mountains rise straight out of the sea, and continue off the coast as a series of islands. Glaciers plunge down from ice fields, and break up into icebergs at the coast.

Below the snowline are evergreen forests of cypress and araucarias (monkey-puzzle trees). On the lower slopes the forests have been burned to create grazing land. Sheep and alpacas are reared for their fine wool, and llamas for use as pack animals. South of the Atacama Desert there are vineyards and orchards, and lush coastal forests where the trees are festooned with lichens, mosses, ferns and trailing vines.

Land of the gauchos

Thorny quebracho trees and patches of woodland are scattered across the grasslands of the chacos. In the rainy summer the land gets wet and swampy, but winter is so dry that the mud cracks up. In the east different crops are grown using irrigation: fruit trees, grapes, sugar cane, cotton, tobacco and maize. In the west there are huge cattle ranches. Beef, cattle hides and wool are major exports from Argentina and Uruguay. Further south, the more fertile pampas are also important cattle country. Cowboys called gauchos round up the cattle and lasso them with ropes. Vast amounts of wheat are grown here, too, for export.

The cold south

Patagonia is a dry, stormy plateau, a cold semi-desert where the wind ships up dust and sand and hurls clouds across the sky. Winter temperatures seldom rise above freezing. Sheep graze on the thin grasses that spring up when the snow melts. Deep gorges provide shelter for the sheep farmers. The windswept Falkland Islands have much higher rainfall. Sheep are reared here, too, and there is good fishing offshore. The tourist industry is growing: people come to watch the local seabirds, penguins and fur seals.

► Cattle ranchers in Argentina travel on horseback. These are the South American 'gauchos', highly skilled horsemen who lasso cattle on the run. The wind-pump is being used to bring up water for the cattle.

World maps

In this section the various regions of the world are illustrated by two types of map. The *physical maps* show the world as it actually appears, detailing physical features such as mountains and rivers. The *political maps* illustrate the boundaries between neighbouring countries.

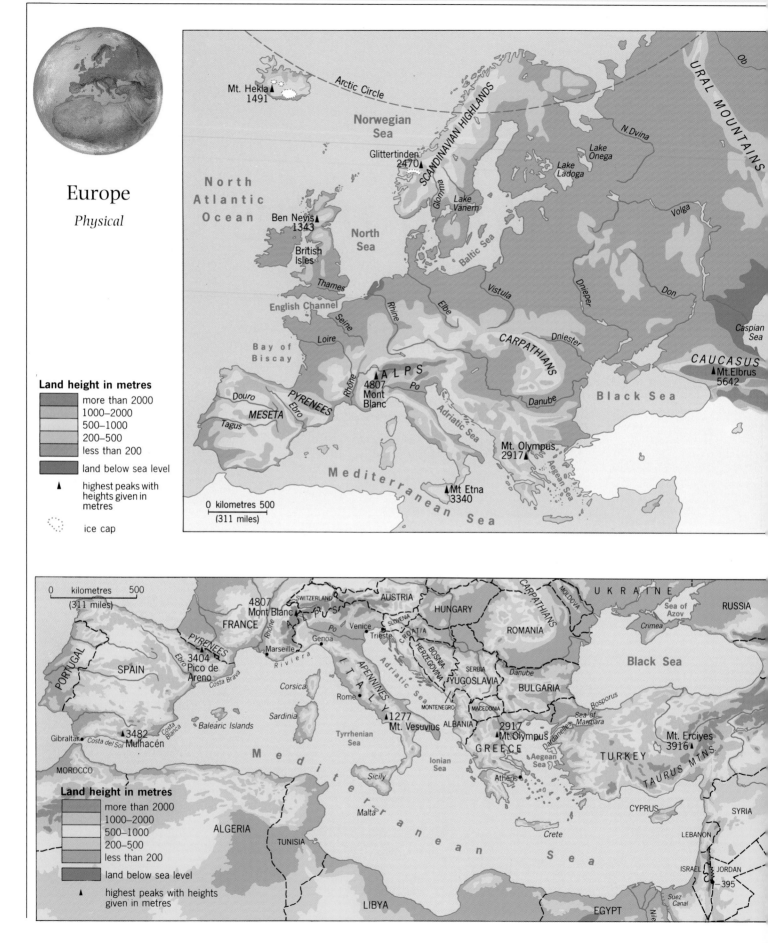

Europe

Physical

Land height in metres

more than 2000
1000–2000
500–1000
200–500
less than 200

land below sea level

▲ highest peaks with heights given in metres

⋰⋱ ice cap

Mt. Hekla ▲
1491

Arctic Circle

Norwegian
Sea

SCANDINAVIAN HIGHLANDS

Glittertinden
2470 ▲

Glomma

North
Atlantic
Ocean

Ben Nevis ▲
1343

British
Isles

North
Sea

Lake
Onega

Lake
Ladoga

Lake
Vänern

Baltic Sea

N. Dvina

Volga

Ob

URAL MOUNTAINS

Thames

English Channel

Seine

Rhine

Elbe

Vistula

Dnieper

Don

Caspian
Sea

Loire

Bay of
Biscay

Douro

MESETA

Tagus

Ebro

PYRENEES

Rhône

ALPS
4807
Mont
Blanc

Po

Danube

Dniester

CARPATHIANS

CAUCASUS
▲ Mt.Elbrus
5642

Black Sea

Adriatic Sea

Mt. Olympus
2917 ▲

Aegean Sea

Mediterranean Sea

▲ Mt Etna
3340

0 kilometres 500
(311 miles)

---

0 kilometres 500
(311 miles)

4807
Mont Blanc

SWITZERLAND

AUSTRIA

HUNGARY

CARPATHIANS

MOLDOVA

UKRAINE

RUSSIA

Sea of
Azov

FRANCE

A L P S

Rhône

Po

Venice

SLOVENIA

CROATIA

ROMANIA

Crimea

PYRENEES
3404
Pico de
Areno

Ebro

Marseille

Genoa

Trieste

Riviera

BOSNA
HERZEGOVINA

SERBIA

Danube

Black Sea

PORTUGAL

SPAIN

Corsica

A P E N N I N E S

Adriatic sea

YUGOSLAVIA

BULGARIA

Bosporus

Sea of
Marmara

Costa Brava

I T A L Y

Rome

MONTENEGRO

MACEDONIA

Mt. Erciyes
3916 ▲

Gibraltar
Costa del Sol

▲3482
Mulhacén

Costa
Blanca

Balearic Islands

Sardinia

Tyrrhenian
Sea

▲1277
Mt. Vesuvius

ALBANIA

2917
▲ Mt.Olympus

GREECE

Dardanelles

Aegean
Sea

TURKEY

TAURUS MTNS.

MOROCCO

Ionian
Sea

Sicily

Athens

CYPRUS

SYRIA

M e d i t e r r a n e a n   S e a

Malta

Crete

LEBANON

ALGERIA

TUNISIA

ISRAEL
JORDAN
▲ 395

Land height in metres

more than 2000
1000–2000
500–1000
200–500
less than 200

land below sea level

▲ highest peaks with heights
given in metres

LIBYA

EGYPT

Suez
Canal

Nile

Europe

Political

Mediterranean

Physical

Asia

Political

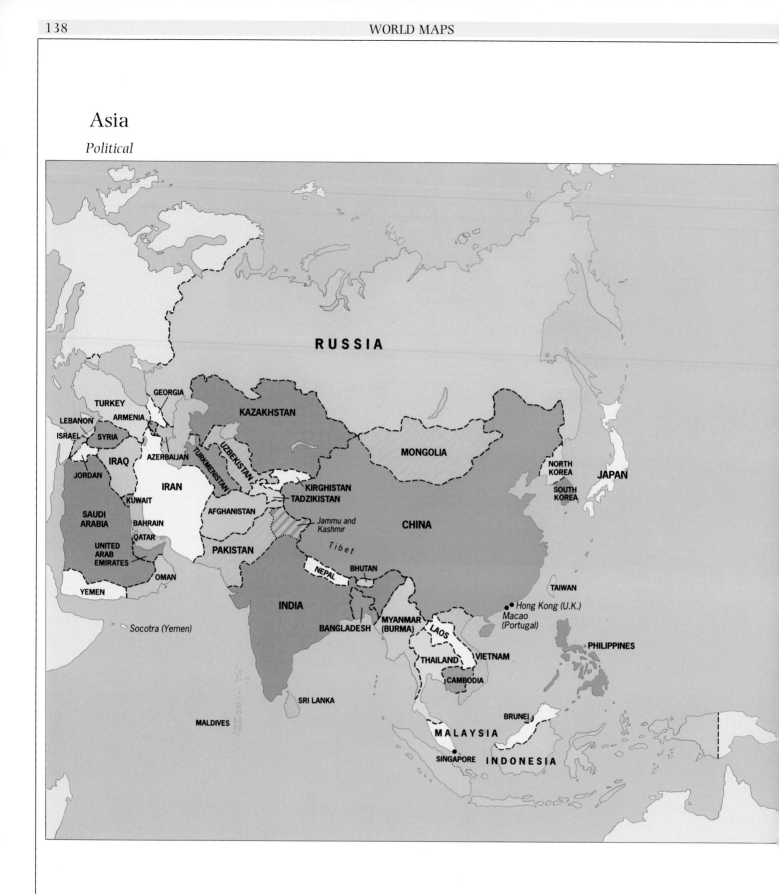

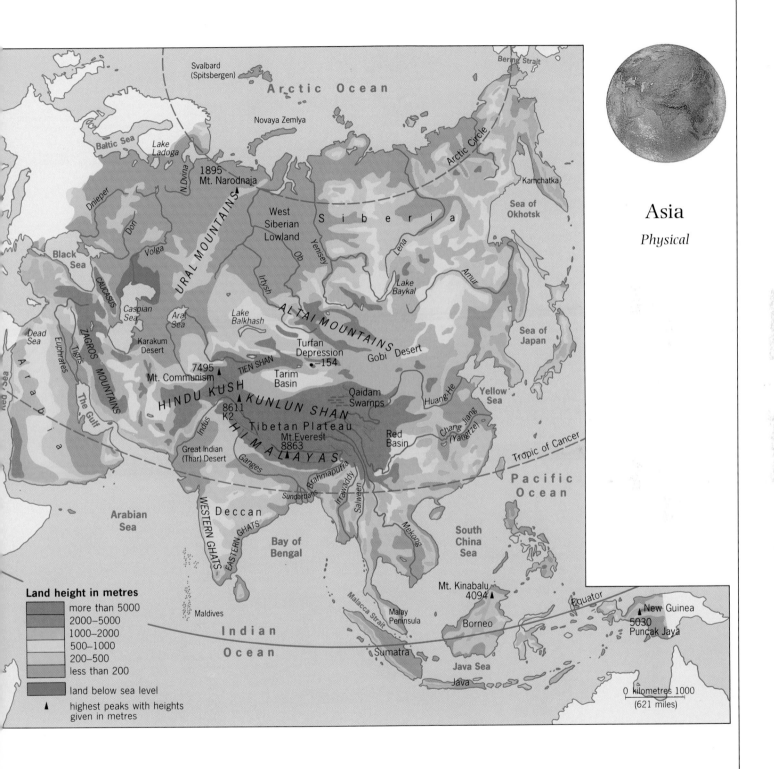

Asia

Physical

Svalbard (Spitsbergen)

Arctic Ocean

Bering Strait

Novaya Zemlya

Baltic Sea

Lake Ladoga

N.Dvina

1895 Mt. Narodnaja

West Siberian Lowland

S i b e r i a

Arctic Circle

Kamchatka

Sea of Okhotsk

Dnieper

Don

Volga

URAL MOUNTAINS

Yenisey

Ob

Lena

Amur

Black Sea

CAUCASUS

Caspian Sea

Aral Sea

Irtysh

Lake Balkhash

ALTAI MOUNTAINS

Lake Baykal

Sea of Japan

Dead Sea

ZAGROS MOUNTAINS

Euphrates

Tigris

Karakum Desert

Turfan Depression

Gobi Desert

A r a b i a

The Gulf

7495 ▲ Mt. Communism

TIEN SHAN

−154

HINDU KUSH

8611 K2

▲ KUNLUN SHAN

Tarim Basin

Qaidam Swamps

Huang He

Yellow Sea

Indus

HIMALAYAS

Tibetan Plateau

Mt.Everest 8863

Red Basin

Chang Jiang (Yangtze)

Great Indian (Thar) Desert

Ganges

Brahmaputra

Tropic of Cancer

Pacific Ocean

Arabian Sea

WESTERN GHATS

Deccan

EASTERN GHATS

Bay of Bengal

Irrawaddy

Salween

Mekong

South China Sea

Maldives

Malacca Strait

Malay Peninsula

Mt. Kinabalu 4094 ▲

Equator

New Guinea

5030 Puncak Jaya

I n d i a n

Borneo

Sumatra

Java Sea

Java

O c e a n

Sunderbans

Land height in metres

- more than 5000
- 2000–5000
- 1000–2000
- 500–1000
- 200–500
- less than 200
- land below sea level
- ▲ highest peaks with heights given in metres

0 kilometres 1000
(621 miles)

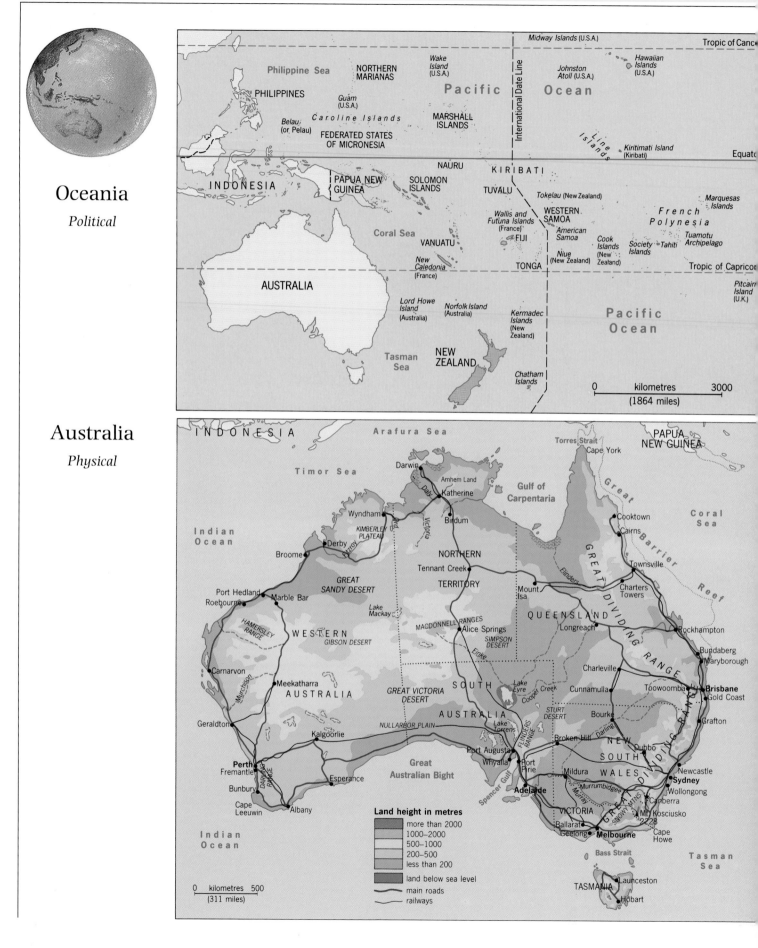

Oceania

Political

Australia

Physical

Oceania Political

Midway Islands (U.S.A.) — Tropic of Cancer

Philippine Sea — NORTHERN MARIANAS — Wake Island (U.S.A.) — Johnston Atoll (U.S.A.) — Hawaiian Islands (U.S.A.)

PHILIPPINES — Pacific Ocean — International Date Line

Guam (U.S.A.) — Caroline Islands — MARSHALL ISLANDS — Line Islands — Kiritimati Island (Kiribati)

Belau (or Pelau) — FEDERATED STATES OF MICRONESIA — Equator

INDONESIA — NAURU — KIRIBATI

PAPUA NEW GUINEA — SOLOMON ISLANDS — TUVALU — Tokelau (New Zealand) — Marquesas Islands

Wallis and Futuna Islands (France) — WESTERN SAMOA — French Polynesia

Coral Sea — VANUATU — FIJI — American Samoa — Cook Islands (New Zealand) — Society Islands — Tahiti — Tuamotu Archipelago

New Caledonia (France) — Niue (New Zealand) — TONGA — Tropic of Capricorn

Lord Howe Island (Australia) — Norfolk Island (Australia) — Kermadec Islands (New Zealand) — Pacific Ocean — Pitcairn Island (U.K.)

AUSTRALIA

Tasman Sea — NEW ZEALAND — Chatham Islands

0 kilometres 3000
(1864 miles)

Australia Physical

INDONESIA — Arafura Sea — Torres Strait — Cape York — PAPUA NEW GUINEA

Timor Sea — Darwin — Arnhem Land — Gulf of Carpentaria — Great — Coral Sea

Katherine — Cooktown — Cairns

Wyndham — Birdum — Townsville

Indian Ocean — KIMBERLEY PLATEAU — Derby — NORTHERN — Charters Towers — Barrier

Broome — Tennant Creek — TERRITORY — Mount Isa — QUEENSLAND — Reef

GREAT SANDY DESERT — MACDONNELL RANGES — Alice Springs — Longreach — Rockhampton

Port Hedland — Marble Bar — Lake Mackay — SIMPSON DESERT — Bundaberg — Maryborough

Roebourne — HAMERSLEY RANGE — WESTERN — GIBSON DESERT — Finke — Charleville — Toowoomba — Brisbane

Carnarvon — Murchison — Meekatharra — AUSTRALIA — GREAT VICTORIA DESERT — SOUTH — Lake Eyre — Cooper Creek — Cunnamulla — Gold Coast

Geraldton — Kalgoorlie — NULLARBOR PLAIN — AUSTRALIA — STURT DESERT — Bourke — Grafton — NEW

Port Augusta — FLINDERS RANGE — Broken Hill — Dubbo — SOUTH

Perth — Fremantle — DARLING RANGE — Esperance — Great Australian Bight — Whyalla — Port Pirie — Mildura — Darling — WALES — Newcastle

Bunbury — Spencer Gulf — Adelaide — Murrumbidgee — Sydney — Wollongong

Cape Leeuwin — Albany — VICTORIA — Murray — SNOWY MTNS — Canberra — Mt Kosciusko 2228 — Cape Howe

Indian Ocean — Ballarat — Geelong — Melbourne — Bass Strait — Tasman Sea

0 kilometres 500
(311 miles)

TASMANIA — Launceston — Hobart

Land height in metres
- more than 2000
- 1000–2000
- 500–1000
- 200–500
- less than 200
- land below sea level
- main roads
- railways

Africa

Political

Madeira
(Portugal)

Canary
Islands
(Spain)

MOROCCO
TUNISIA
WESTERN
SAHARA
ALGERIA
LIBYA
EGYPT

CAPE
VERDE

MAURITANIA
MALI
NIGER
CHAD
SUDAN
ERITREA

SENEGAL
THE GAMBIA
GUINEA
BISSAU
GUINEA
SIERRA
LEONE
BURKINA
DJIBOUTI
ETHIOPIA

LIBERIA
COTE
D'IVOIRE
GHANA
BENIN
TOGO
NIGERIA
CENTRAL AFRICAN
REPUBLIC
SOMALIA

EQUATORIAL
GUINEA
CAMEROON
UGANDA
KENYA

SÃO TOME
AND
PRINCIPE
GABON
CONGO
ZAÏRE
RWANDA
BURUNDI
TANZANIA

COMOROS

ANGOLA
ZAMBIA
MALAWI
MADAGASCAR

MAURITIUS

NAMIBIA
ZIMBABWE
MOZAMBIQUE
Réunion
(France)

BOTSWANA
SWAZILAND

SOUTH
AFRICA
LESOTHO

Africa

Physical

North
Atlantic
Ocean

Mediterranean Sea

Jebel Toubkal
4165
ATLAS MOUNTAINS

−134
Qattara
Depression

Tropic of Cancer

SAHARA   DESERT

Lake
Nasser

Senegal
Gambia
Niger

Lake
Chad

Chari

Nile
Red Sea

White Nile
Blue Nile

Lake
Assal

ETHIOPIAN
HIGHLANDS

Benue
Lake
Volta

Sudd
Swamp

Equator
Gulf of Guinea

Oubangui
Zaïre

RIFT VALLEY

Congo
Basin

RIFT VALLEY

Mt. Kenya
5199

Lake
Victoria

South
Atlantic
Ocean

Zaïre
Kasai

Rift
Valley
Mt. Kilimanjaro
5895

Lualaba

Lake
Tanganyika

Indian
Ocean

Lake Nyasa
(Lake Malawi)

2876
Maromokotro

Cubango

Zambezi

0   kilometres 1000

(621 miles)

Cunene

Okavango
Swamp

Mozambique Channel

Limpopo

NAMIB DESERT
Tropic of Capricorn
KALAHARI
DESERT

Land height in metres

more than 2000
1000–2000
500–1000
200–500
less than 200

land below sea level

▲ highest peaks with
heights given in metres

Orange
Vaal
DRAKENSBERG
3482
Thabana-Ntlenyana

Cape of
Good Hope

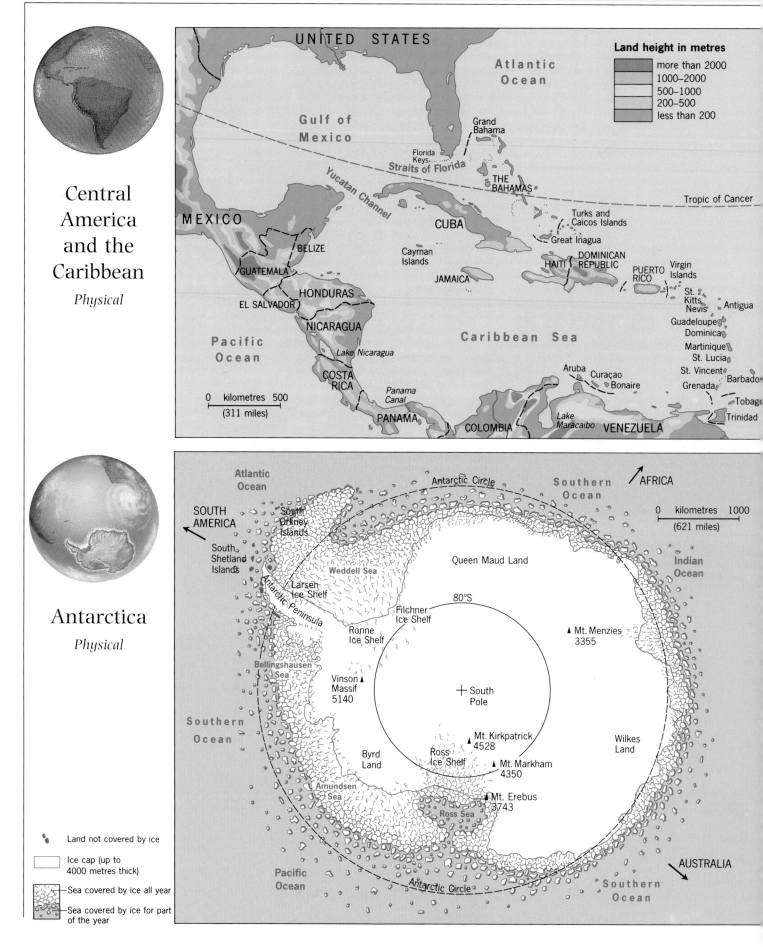

Central America and the Caribbean

Physical

Antarctica

Physical

Land height in metres

more than 2000
1000–2000
500–1000
200–500
less than 200

UNITED STATES

Atlantic Ocean

Gulf of Mexico

Grand Bahama

Florida Keys

Straits of Florida

THE BAHAMAS

Tropic of Cancer

Yucatan Channel

MEXICO

CUBA

Turks and Caicos Islands

Great Inagua

BELIZE

Cayman Islands

DOMINICAN REPUBLIC

HAITI

PUERTO RICO

Virgin Islands

GUATEMALA

JAMAICA

St. Kitts
Nevis

Antigua

HONDURAS

Guadeloupe
Dominica

EL SALVADOR

NICARAGUA

Caribbean Sea

Martinique
St. Lucia

Pacific Ocean

Lake Nicaragua

St. Vincent

Barbados

0   kilometres   500

(311 miles)

COSTA RICA

Panama Canal

Aruba

Curaçao

Bonaire

Grenada

Tobago

PANAMA

COLOMBIA

Lake Maracaibo

VENEZUELA

Trinidad

Atlantic Ocean

Antarctic Circle

Southern Ocean

AFRICA

SOUTH AMERICA

South Orkney Islands

0   kilometres   1000

(621 miles)

South Shetland Islands

Weddell Sea

Queen Maud Land

Indian Ocean

Larsen Ice Shelf

Antarctic Peninsula

Filchner Ice Shelf

80°S

▲ Mt. Menzies
3355

Ronne Ice Shelf

Bellingshausen Sea

Vinson ▲ Massif
5140

+ South Pole

Wilkes Land

Southern Ocean

Byrd Land

Ross Ice Shelf

▲ Mt. Kirkpatrick
4528

Amundsen Sea

▲ Mt. Markham
4350

▲ Mt. Erebus
3743

Ross Sea

AUSTRALIA

Pacific Ocean

Antarctic Circle

Southern Ocean

Land not covered by ice

Ice cap (up to 4000 metres thick)

Sea covered by ice all year

Sea covered by ice for part of the year

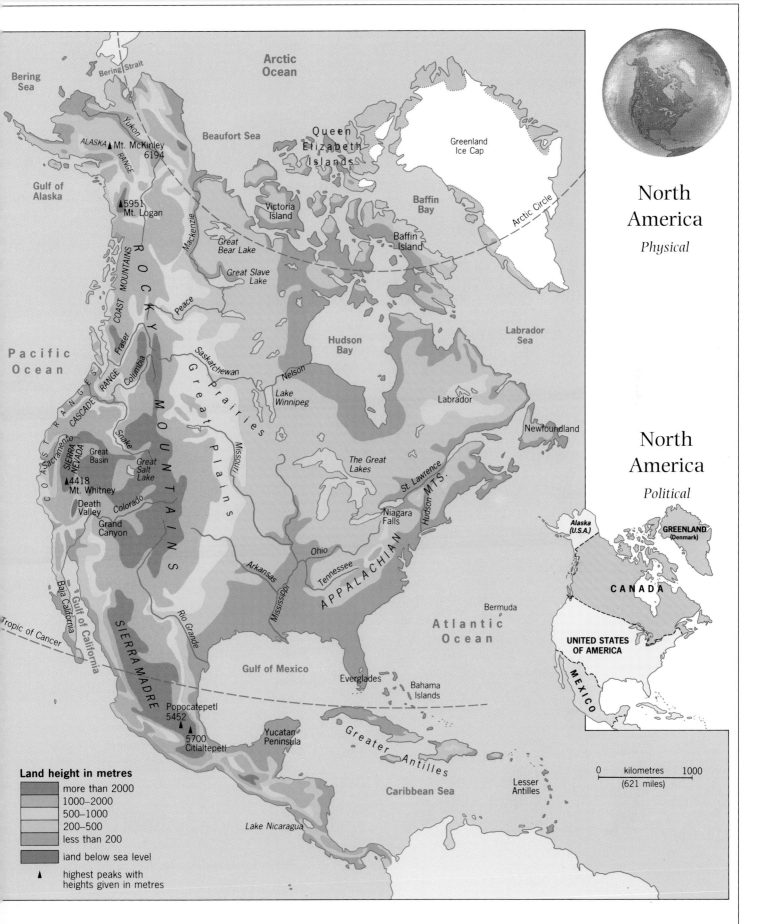

North America
Physical

North America
Political

Arctic Ocean

Bering Strait

Bering Sea

ALASKA ▲ Mt. McKinley 6194

Beaufort Sea

Queen Elizabeth Islands

Greenland Ice Cap

Arctic Circle

Gulf of Alaska

Yukon

RANGE

▲ 5951 Mt. Logan

Victoria Island

Great Bear Lake

Baffin Bay

Baffin Island

ROCKY

Mackenzie

Great Slave Lake

COAST MOUNTAINS

Peace

Pacific Ocean

Fraser

Columbia

Saskatchewan

Great Prairies

Nelson

Hudson Bay

Labrador Sea

Labrador

Newfoundland

RANGES

CASCADE RANGE

MOUNTAINS

Great Plains

Lake Winnipeg

Snake

Sacramento

SIERRA NEVADA

Great Basin

Great Salt Lake

Missouri

The Great Lakes

St. Lawrence

Niagara Falls

Hudson

APPALACHIAN MTS.

COAST

▲ 4418 Mt. Whitney

Death Valley

Colorado

Grand Canyon

Arkansas

Ohio

Tennessee

Mississippi

Tropic of Cancer

Gulf of California

Baja California

Rio Grande

SIERRA MADRE

Bermuda

Atlantic Ocean

Gulf of Mexico

Everglades

Bahama Islands

Popocatepetl 5452 ▲

▲ 5700 Citlaltepetl

Yucatan Peninsula

Greater Antilles

Lesser Antilles

Caribbean Sea

Lake Nicaragua

Alaska (U.S.A.)

GREENLAND (Denmark)

CANADA

UNITED STATES OF AMERICA

MEXICO

0   kilometres   1000
(621 miles)

Land height in metres

- more than 2000
- 1000–2000
- 500–1000
- 200–500
- less than 200
- land below sea level
- ▲ highest peaks with heights given in metres

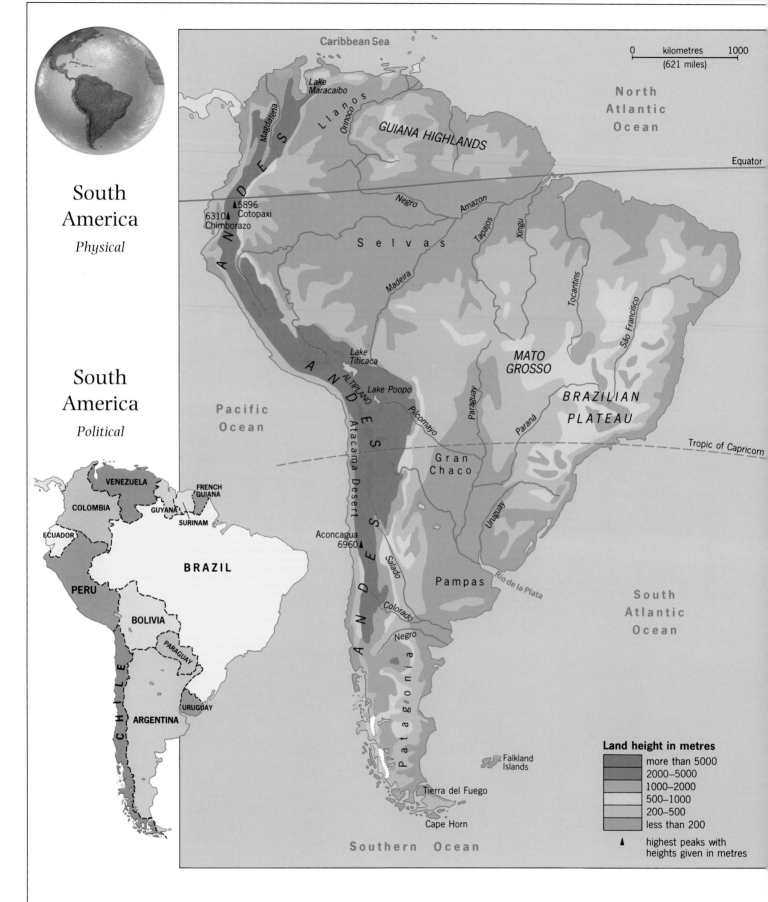

South
America

Physical

South
America

Political

0   kilometres   1000
(621 miles)

Caribbean Sea

Lake
Maracaibo

Magdalena

Orinoco

Llanos

GUIANA HIGHLANDS

North
Atlantic
Ocean

Equator

A
N
D
E
S

▲5896
Cotopaxi
6310▲
Chimborazo

Negro

Amazon

S e l v a s

Tapajos

Xingu

Madeira

Tocantins

São Francisco

Lake
Titicaca

ALTIPLANO

Lake Poopo

Pilcomayo

Paraguay

MATO
GROSSO

BRAZILIAN

PLATEAU

Pacific
Ocean

A
N
D
E
S

Atacama Desert

Paraná

Tropic of Capricorn

G r a n
C h a c o

Uruguay

Aconcagua
6960▲

Salado

Pampas

Rio de la Plata

South
Atlantic
Ocean

Colorado

Negro

A
N
D
E
S

P
a
t
a
g
o
n
i
a

Falkland
Islands

Tierra del Fuego

Cape Horn

Southern   Ocean

VENEZUELA

FRENCH
GUIANA

COLOMBIA

GUYANA

SURINAM

ECUADOR

BRAZIL

PERU

BOLIVIA

PARAGUAY

C
H
I
L
E

URUGUAY

ARGENTINA

Land height in metres

more than 5000
2000–5000
1000–2000
500–1000
200–500
less than 200

▲   highest peaks with
     heights given in metres

Countries at a glance

All the countries listed in these tables are independent except those printed in *italic* type. The populations are 1991 estimates, with the exception of the United States (1990 census) and the non-independent countries, which are the latest available estimates.

The languages are the official languages plus, in some cases, others commonly spoken. But throughout the world people speak hundreds of different languages. There are also hundreds of traditional religions.

NORTHERN EUROPE

Country	Area	Capital	Population	Language	Religion	Currency
Belgium	30,513 sq km 11,783 sq miles	Brussels	9,946,000	Dutch, French	Christianity	Belgian franc
Denmark	43,007 sq km 16,632 sq miles	Copenhagen	5,122,000	Danish	Christianity	Krone
Estonia	45,100 sq km 17,413 sq miles	Tallinn	1,595,000	Estonian, Russian	Christianity	Kroon
Finland	338,145 sq km 130,559 sq miles	Helsinki (Helsingfors)	4,976,000	Finnish, Swedish	Christianity	Markka
Iceland	103,000 sq km 39,800 sq miles	Reykjavik	255,000	Icelandic	Christianity	Krona
Ireland, Republic of	70,283 sq km 27,137 sq miles	Dublin	3,755,000	English, Irish Gaelic	Christianity	Punt
Latvia	63,700 sq km 24,595 sq miles	Riga	2,718,000	Latvian, Russian	Christianity	Lat
Lithuania	65,200 sq km 25,114 sq miles	Vilnius	3,742,000	Lithuanian, Russian	Christianity	Litas
Netherlands	41,863 sq km 16,163 sq miles	Amsterdam	14,803,000	Dutch	Christianity	Guilder
Norway	386,958 sq km 149,407 sq miles	Oslo	4,224,000	Norwegian	Christianity	Krone
Sweden	440,945 sq km 173,732 sq miles	Stockholm	8,336,000	Swedish	Christianity	Krona
United Kingdom	244,100 sq km 94,248 sq miles	London	57,537,000	English, Welsh	Christianity, Islam, others	Pound

CENTRAL AND EASTERN EUROPE

Country	Area	Capital	Population	Language	Religion	Currency
Austria	83,849 sq km 32,376 sq miles	Vienna	7,490,000	German	Christianity	Schilling
Czech Republic	78,868 sq km 30,451 sq miles	Prague	10,700,000	Czech	Christianity, Islam	Koruna
France	543,965 sq km 212,935 sq miles	Paris	56,375,000	French	Christianity, Islam	French franc
Germany	356,829 sq km 137,804 sq miles	Berlin	77,454,000	German	Christianity	Mark
Hungary	93,032 sq km 39,920 sq miles	Budapest	10,554,000	Magyar	Christianity	Forint
Liechtenstein	160 sq km 62 sq miles	Vaduz	28,000	German	Christianity	Swiss franc
Luxembourg	2,586 sq km 998 sq miles	Luxembourg	367,000	French, German, Letzeburgesch	Christianity	Franc
Poland	312,683 sq km 120,725 sq miles	Warsaw	38,611,000	Polish	Christianity	Zloty
Romania	237,500 sq km 91,700 sq miles	Bucharest	23,379,000	Romanian	Christianity	Leu
Slovak Republic	49,008 sq km 18,922 sq miles	Bratislava	5,000,000	Slovak	Christianity	Koruna
Switzerland	41,293 sq km 15,943 sq miles	Bern	6,528,000	German, French, Italian, Romansh	Christianity	Swiss franc

SOUTHERN EUROPE

Country	Area	Capital	Population	Language	Religion	Currency
Albania	28,748 sq km 11,100 sq miles	Tiranë	3,298,000	Albanian	Islam, Christianity	Lek
Andorra	363 sq km 175 sq miles	Andorra	52,000	Catalan	Christianity	French franc and Spanish peseta
Bosnia - Herzegovina	51,129 sq km 19,741 sq miles	Sarajevo	4,116,000	Croatian	Christianity, Islam	Dinar
Bulgaria	110,912 sq km 42,823 sq miles	Sofia	9,015,000	Bulgarian	Christianity, Islam	Lev
Croatia	56,537 sq km 21,829 sq miles	Zagreb	4,601,000	Serbo-Croat	Christianity	Dinar
Greece	131,990 sq km 50,962 sq miles	Athens	10,062,000	Greek	Christianity	Drachma
Italy	301,268 sq km 116,320 sq miles	Rome	57,838,000	Italian	Christianity	Italian lira
Macedonia (former Yugoslavian Republic)	25,713 sq km 9,156 sq miles	Skopje	2,000,000	Macedonian	Christianity, Islam	Denar
Malta	316 sq km 122 sq miles	Valletta	354,000	Maltese, English	Christianity	Maltese lira
Monaco	1·90 sq km 0·68 sq miles	Monaco	30,000	French	Christianity	French franc
Portugal	88,941 sq km 34,340 sq miles	Lisbon	10,314,000	Portuguese	Christianity	Escudo
San Marino	61 sq km 24 sq miles	San Marino	23,000	Italian	Christianity	Italian lira
Slovenia	20,251 sq km 7,819 sq miles	Ljubljana	1,892,000	Slovenian	Christianity	Dinar
Spain	504,750 sq km 194,885 sq miles	Madrid	39,479,000	Spanish	Christianity	Peseta
Vatican City State	0·44 sq km 0·17 sq miles	Vatican City	1,000	Italian, Latin	Christianity	Italian lira
Yugoslavia (former)	102,172 sq km 39,449 sq miles	Belgrade	9,989,000	Serbo-Croat	Christianity	Dinar

NOTES: European Turkey is in under Middle East. Former Yugoslavia consists of the republics of Serbia and Montenegro

RUSSIA AND ITS NEIGHBOURS

Country	Area	Capital	Population	Language	Religion	Currency
Belarus	207,600 sq km 80,155 sq miles	Minsk	9,878,000	Belarussian, Russian	Christianity, Islam	Rouble
Georgia	69,700 sq km 26,911 sq miles	Tbilisi	5,167,000	Georgian, Russian	Christianity	Rouble
Moldova	33,700 sq km 13,012 sq miles	Chisinau	4,080,000	Romanian, Moldovan, Russian	Christianity	Rouble
Russia (inc. Asian Russia)	17,075,000 sq km 6,582,695 sq miles	Moscow	142,117,000	Russian	Christianity	Rouble
Ukraine	603,700 sq km 233,090 sq miles	Kiev	50,667,000	Ukrainian, Russian	Christianity	Rouble

CENTRAL ASIA

Country	Area	Capital	Population	Language	Religion	Currency
Afghanistan	652,090 sq km 251,773 sq miles	Kabul	17,666,000	Pushto, Dari	Islam	Afghani
Armenia	30,000 sq km 11,583 sq miles	Yerevan	3,000,000	Armenian, Russian	Christianity	Rouble
Azerbaijan	86,600 sq km 33,440 sq miles	Baku	6,506,000	Azerbaijani, Russian	Islam	Manat
Kazakhstan	2,717,300 sq km 1,049,155 sq miles	Alma Alta	15,648,000	Kazakh, Russian	Islam, Christianity	Rouble
Kirghistan	198,500 sq km 76,641 sq miles	Bishkek	3,886,000	Kirghiz, Russian	Islam, Christianity	Rouble
Tadzikistan	143,100 sq km 55,251 sq miles	Dushanbe	4,365,000	Tadzik, Russian	Islam	Rouble
Turkmenistan	488,100 sq km 188,846 sq miles	Ashkhabad	3,188,000	Turkmen, Russian	Islam	Rouble
Uzbekistan	447,400 sq km 172,742 sq miles	Tashkent	17,498,000	Uzbek, Russian, Tadzik	Islam	Rouble

MIDDLE EAST

Country	Area	Capital	Population	Language	Religion	Currency
Bahrain	688 sq km 265 sq miles	Manama	531,000	Arabic	Islam	Dinar
Cyprus	9,251 sq km 3,572 sq miles	Nicosia	708,000	Greek,	Christianity, Islam	Cyprus pound
Iran	1,648,000 sq km 636,300 sq miles	Teheran	58,073,000	Farsi	Islam	Rial
Iraq	438,317 sq km 169,235 sq miles	Baghdad	18,600,000	Arabic	Islam	Dinar
Israel	20,770 sq km 8,019 sq miles	Jerusalem	4,647,000	Hebrew, Arabic	Judaism, Islam	Shekel
Jordan	91,880 sq km 35,475 sq miles	Amman	3,162,000	Arabic	Islam	Dinar
Kuwait	17,818 sq miles 6,880 sq miles	Kuwait City	2,154,000	Arabic	Islam	Dinar
Lebanon	10,4300 sq km 4,015 sq miles	Beirut	3,026,000	Arabic	Islam	Lebanese pound
Oman	212,457 sq km 82,030 sq miles	Muscat	1,517,000	Arabic	Islam	Rial
Qatar	11,437 sq km 4,416 sq miles	Doha	380,000	Arabic	Islam	Riyal
Saudi Arabia	2,149,690 sq km 830,000 sq miles	Riyadh	13,366,000	Arabic	Islam	Saudi riyal
Syria	185,180 sq km 71,498 sq miles	Damascus	12,941,000	Arabic	Islam	Syrian pound
Turkey	780,576 sq km 188,456 sq miles	Ankara	57,301,000	Turkish	Islam	Lira
United Arab Emirates	83,600 sq km 32,278 sq miles	Abu Dhabi	1,624,000	Arabic	Islam	Dirham
Yemen	528,038 sq km 203,887 sq miles	Sana	10,848,000	Arabic	Islam	Yemeni dinar, Rial

	Country	Area	Capital	Population	Language	Religion	Currency
INDIAN SUBCONTINENT	**Bangladesh**	143,998 sq km 55,598 sq miles	Dhaka	118,702,000	Bengali	Islam, Hinduism	Taka
	Butan	46,500 sq km 17,950 sq miles	Thimpu	1,550,000	Dzongkha, Nepali	Buddhism, Hinduism	Ngultrum
	India	3,287,263 sq km 1,269,219 sq miles	New Delhi	871,208,000	Hindi, English	Hinduism, Islam	Rupee
	Maldives	298 sq km 115 sq miles	Male	227,000	Divehi	Islam	Rufiyaa
	Nepal	147,181 sq km 56,827 sq miles	Kathmandu	19,591,000	Nepali	Hinduism, Buddhism	Rupee
	Pakistan	796,095 sq km 307,374 sq miles	Islamabad	122,666,000	Urdu	Islam	Rupee
	Sri Lanka	65,610 sq km 25,333 sq miles	Colombo	17,639,000	Sinhalese, Tamil	Buddhism Hinduism, Islam	Rupee
FAR EAST	**China**	9,572,678 sq km 3,696,032 sq miles	Beijing	1,115,883,000	Mandarin	Taoism, Buddhism Christianity	Yuan
	Hong Kong	2,916 sq km 1,126 sq miles	Victoria	5,756,000	English, Chinese	Buddhism, Taoism	Hong Kong dollar
	Japan	337,801 sq km 145,870 sq miles	Tokyo	124,025,000	Japanese	Shinto, Buddhism	Yen
	Korea, North	120,538 sq km 46,540 sq miles	Pyongyang	23,432,000	Korean	Ch'ondogyo, traditional beliefs	Won
	Korea, South	99,106 sq km 36,625 sq miles	Seoul	44,018,000	Korean	Buddhism, Christianity	Won
	Mongolia	1,565,000 sq km 604,250 sq miles	Ulan Bator	2,295,000	Mongolian	Buddhism	Tughrik
	Taiwan	36,000 sq km 13,900 sq miles	Taipei	20,923,000	Chinese	Taoism, Buddhism Christianity	New Taiwan dollar

NOTE: Hong Kong is a British dependency, but by a treaty signed in 1984 it is to be transferred to Chinese control

	Country	Area	Capital	Population	Language	Religion	Currency
SOUTH-EAST ASIA	**Brunei**	5,765 sq km 2,226 sq miles	Bandar Seri Begawan	299,000	Malay	Islam,Christianity Buddhism	Brunei dollar
	Cambodia	181,035 sq km 69,898 sq miles	Phnom Penh	7,147,000	Khmer	Buddhism, Islam	Riel
	Indonesia	1,919,443 sq km 741,101 sq miles	Jakarta	183,258,000	Bahasa Indonesian	Islam, Christianity	Rupiah
	Laos	236,800 sq km 91,430 sq miles	Vientiane	4,167,000	Lao	Buddhism	Kip
	Malaysia	329,749 sq km 127,317 sq miles	Kuala Lumpur	17,689,000	Malay	Islam, Buddhism	Malaysian dollar
	Myanmar (Burma)	676,552 sq km 261,218 sq miles	Yangon (Rangoon)	42,146,000	Burmese	Buddhism, Christianity, Islam	Kyat Main
	Philippines	300,000 sq km 116,000 sq miles	Manila	63,826,000	Filipino, English	Christianity, Islam	Peso
	Singapore	618 sq km 239 sq miles	Singapore City	2,728,000	Chinese, English,	Buddhism, Hinduism, Christianity, Taoism	Singapore dollar
	Thailand	513,115 sq km 198,115 sq miles	Bangkok	56,454,000	Thai	Buddhism, Islam	Baht
	Vietnam	329,556 sq km 127,242 sq miles	Hanoi	67,290,000	Vietnamese	Taoism, Buddhism Christianity	Dong

AUSTRALIA, NEW GUINEA, NEW ZEALAND

Country	Area	Capital	Population	Language	Religion	Currency
Australia	7,682,000 sq km 2,966,039 sq miles	Canberra	16,930,000	English	Christianity	Australian dollar
New Zealand	269,057 sq km 103,883 sq miles	Wellington	3,404,000	English	Christianity	New Zealand dollar
Papua New Guinea	462,840 sq km 178,704 sq miles	Port Moresby	4,112,000	English, Pidgin	Traditional beliefs, Christianity	Kina

NOTE: The western half of New Guinea island is part of Indonesia (see South-east Asia)

PACIFIC ISLANDS

Country	Area	Capital	Population	Language	Religion	Currency
Fiji	18,274 sq km 7,056 sq miles	Suva	781,000	English	Christianity, Hinduism, Islam	Fiji dollar
Kiribati	717 sq km 227 sq miles	Tarawa	71,000	Gilbertese, English	Christianity	Australian dollar
Nauru	21 sq km 8 sq miles	None	9,000	Nauruan, English	Christianity	Australian dollar
Solomon Islands	27,556 sq km 10,639 sq miles	Honiara	340,000	English, Pidgin	Christianity	Solomon Is. dollar
Tonga	748 sq km 289 sq miles	Nuku'alofa	98,000	Tongan	Christianity	Pa'anga
Tuvalu	26 sq km 10 sq miles	Funafuti	9,000	Tuvaluan	Christianity	Australian dollar
Vanuatu	12,189 sq km 4,706 sq miles	Port-Vila	12,189	Bislama, English, French	Christianity	Vatu
Western Samoa	2,831 sq km 1,093 sq miles	Apia	177,000	Samoan, English	Christianity	Tala

NOTES: There are between 20,000 and 30,000 islands in the Pacific Ocean. Most of them are states, provinces, or dependencies of Australia, Chile, France, Indonesia, New Zealand, the United Kingdom or the United States.

TEMPERATE SOUTH AMERICA

Country	Area	Capital	Population	Language	Religion	Currency
Argentina	2,780,092 sq km 1,073,400 sq miles	Buenos Aires	32,700,000	Spanish	Christianity	Austral
Chile	756,945 sq km 292,258 sq miles	Santiago	13,377,000	Spanish	Christianity	Peso
Paraguay	406,752 sq km 157,048 sq miles	Asuncion	4,392,000	Spanish	Christianity	Guarani
Uruguay	177,414 sq km 68,500 sq miles	Montevideo	3,151,000	Spanish	Christianity	New Uruguyan peso

CENTRAL AMERICA AND THE CARIBBEAN

Country	Area	Capital	Population	Language	Religion	Currency
Antigua and Barbuda	442 sq km 171 sq miles	St John's	82,000	English	Christianity	East Caribbean dollar
Bahamas	13,878 sq km 5,358 sq miles	Nassau	256,000	English	Christianity	Bahamisan dollar
Barbados	431 sq km 166 sq miles	Bridgetown	263,000	English	Christianity	Barbados dollar
Belize	22,965 sq km 8,867 sq miles	Belmopan	184,000	English, Spanish	Christianity	Belizean dollar
Costa Rica	51,100 sq km 19,730 sq miles	San José	3,083,000	Spanish	Christianity	Colón
Cuba	119,861 sq km 46,279 sq miles	Havana	10,415,000	Spanish	Christianity	Peso
Dominica	751 sq km 290 sq miles	Roseau	86,000	English	Christianity	East Caribbean dollar
Dominican Republic	48,734 sq km 18,816 sq miles	Santo Domingo	7,312,000	Spanish	Christianity	Peso
El Salvador	21,041 sq km 8,124 sq miles	San Salvador	5,381,000	Spanish	Christianity	Colón
Grenada	344 sq km 133 sq miles	St George's	92,000	English	Christianity	East Caribbean dollar
Guatemala	108,889 sq km 42,042 sq miles	Guatemala City	9,462,000	Spanish	Christianity	Quetzal
Haiti	27,750 sq km 10,714 sq miles	Port-au-Prince	6,427,000	French	Christianity, Voodoo	Gourde
Honduras	112,088 sq km 43,277 sq miles	Tegucigalpa	5,292,000	Spanish	Christianity	Lempira
Jamaica	10,991 sq km 4,244 sq miles	Kingston	2,557,000	English	Christianity	Jamaican dollar
Mexico	1,958,201 sq km 756,066 sq miles	Mexico City	90,379,000	Spanish	Christianity	Peso
Nicaragua	130,000 sq km 50,193 sq miles	Managua	3,994,000	Spanish	Christianity	Córdoba
Panama	78,200 sq km 30,193 sq miles	Panama City	2,464,000	Spanish	Christianity	Balboa
Puerto Rico	9,103 sq km 3,515 sq miles	San Juan	3,282,000	Spanish, English	Christianity	US dollar
St Christopher and Nevis	261 sq km 101 sq miles	Basseterre	48,000	English	Christianity	East Caribbean dollar
St Lucia	616 sq km 240 sq miles	Castries	157,000	English	Christianity	East Caribbean dollar
St Vincent and the Grenadines	388 sq km 150 sq miles	Kingstown	115,000	English	Christianity	East Caribbean dollar
Trinidad and Tobago	5,128 sq km 1,980 sq miles	Port-of-Spain	1,303,000	English, French	Christianity	East Caribbean dollar

NOTE: Puerto Rico is a United States commonwealth.

TROPICAL SOUTH AMERICA

Country	Area	Capital	Population	Language	Religion	Currency
Bolivia	1,098,581 sq km 424,165 sq miles	La Paz	7,520,000	Spanish, Aymara, Quechua	Christianity	Boliviano
Brazil	8,482,081 sq km 3,274,950 sq miles	Brasilia	153,180,000	Portuguese	Christianity	Cruziero
Colombia	1,141,748 sq km 440,831 sq miles	Bogota	32,414,000	Spanish	Christianity	Peso
Ecuador	283,561 sq km 109,484 sq miles	Quito	11,069,000	Spanish	Christianity	Sucre
French Guiana	91,000 sq km 35,135 sq miles	Cayenne	73,000	French	Christianity	French franc
Guyana	214,969 sq km 83,000 sq miles	Georgetown	881,000	English	Christianity, Hinduism, Islam	Guyana dollar
Peru	1,285,216 sq km 496,225 sq miles	Lima	22,857,000	Spanish, Quechua	Christianity	Inti
Surinam	163,265 sq km 63,037 sq miles	Paramaribo	409,000	Dutch	Christianity, Hinduism, Islam	Surinam guilder
Venezuela	912,050 sq km 352,145 sq miles	Caracas	20,202,000	Spanish	Christianity	Bolivar

NOTE: French Guiana is the smallest of the three Guianas, and the only one not independent. It is administered as an overseas department of France.

NORTH AFRICA

Country	Area	Capital	Population	Language	Religion	Currency
Algeria	2,381,741 sq km 919,595 sq miles	Algiers	26,097,000	Arabic	Islam	Dinar
Chad	1,284,000 sq km 495,755 sq miles	N'Djamena	5,822,000	Arabic, French	Islam, Christianity	Franc CFA
Djibouti	23,200 sq km 8,958 sq miles	Djibouti	418,000	French	Islam	Djibouti franc
Egypt (Inc. Asian Egypt)	1,001,449 sq km 386,662 sq miles	Cairo	54,673,000	Arabic	Islam	Egyptian pound
Ethiopia	1,221,900 sq km 471,778 sq miles	Addis Ababa	49,263,000	Amharic	Christianity, Islam	Birr
Libya	1,759,540 sq km 679,362 sq miles	Tripoli	4,780,000	Arabic	Islam	Dinar
Mali	1,240,192 sq km 478,841 sq miles	Bamoko	8,562,000	French	Islam, traditional beliefs	Franc CFA
Mauritania	1,030,700 sq km 397,956 sq miles	Nouakchott	2,081,000	Arabic, French	Islam	Ouguiya
Morocco	458,730 sq km 177,117 sq miles	Rabat	25,730,000	Arabic	Islam	Dirham
Niger	1,267,000 sq km 489,200 sq miles	Niamey	7,952,000	French, Hausa	Islam, traditional beliefs	Franc CFA
Somalia	637,657 sq km 246,201 sq miles	Mogadishu	7,734,000	Somali, Arabic	Islam	Somali shilling
Sudan	2,505,813 sq km 967,500 sq miles	Khartoum	25,923,000	Arabic	Islam, traditional beliefs	Sudanese pound
Tunisia	163,610 sq km 63,170 sq miles	Tunis	8,331,000	Arabic	Islam	Dinar
Western Sahara	266,000 sq km 102,703 sq miles	None	180,000	Arabic	Islam	Peseta

NOTE: Western Sahara is a former Spanish province, claimed by Morocco. The claim is disputed by some of its mainly nomadic people.

CENTRAL AFRICA

Country	Area	Capital	Population	Language	Religion	Currency
Benin	112,622 sq km 43,484 sq miles	Porto-Novo	4,895,000	French	Christianity, Islam, traditional beliefs	Franc CFA
Burkina	274,200 sq km 105,869 sq miles	Ougadougou	9,263,000	French	Christianity, Islam, traditional beliefs	Franc CFA
Burundi	27,834 sq km 10,747 sq miles	Bujumbura	5,609,000	Kirundi, French	Christianity, traditional beliefs	Burundi franc
Cameroon	475,442 sq km 240,535 sq miles	Yaoundé	11,550,000	French, English	Christianity, traditional beliefs	Franc CFA
Cape Verde	4,033 sq km 1,557 sq miles	Praia	391,000	Portuguese	Christianity	Escudo
Central African Republic	622,984 sq km 240,535 sq miles	Bangui	2,987,000	French	Traditional beliefs, Christianity, Islam	Franc CFA
Congo	342,000 sq km 132,047 sq miles	Brazzaville	2,197,000	French	Traditional beliefs, Christianity	Franc CFA
Côte d'Ivoire (Ivory Coast)	322,463 sq km 124,504 sq miles	Abidjan	13,089,000	French	Traditional beliefs, Islam, Christianity	Franc CFA
Equatorial Guinea	28,051 sq km 10,831 sq miles	Malabo	451,000	Fang, Spanish	Christianity, traditional beliefs	Franc CFA
Gabon	267,667 sq km 103,347 sq miles	Libreville	1,210,000	French, local tongues	Christianity	Franc CFA
Gambia	11,295 sq km 4,361 sq miles	Banjul	881,000	English	Islam, Christianity	Dalasi
Ghana	238,537 sq km 92,100 sq miles	Accra	15,486,000	English	Christianity, traditional beliefs	Cedi
Guinea	245,857 sq km 94,926 sq miles	Conakry	7,051,000	French	Islam, traditional beliefs	Guinea franc
Guinea-Bissau	36,125 sq km 13,948 sq miles	Bissau	1,009,000	Portuguese	Islam, traditional beliefs	Peso
Kenya	580,367 sq km 224,081 sq miles	Nairobi	26,160,000	Swahili, English	Christianity, Islam, traditional beliefs	Kenya shilling
Liberia	111,370 sq km 43,000 sq miles	Monrovia	2,637,000	English	Christianity, Islam, traditional beliefs	Liberian dollar
Nigeria	923,768 sq km 356,669 sq miles	Lagos	116,926,000	English	Islam, Christianity	Naira
Rwanda	26,338 sq km 10,169 sq miles	Kigali	7,479,000	Kinyarwanda, French	Christianity, traditional beliefs	Rwanda franc
São Tomé and Principe	964 sq km 372 sq miles	Sao Tomé	128,000	Portuguese	Christianity	Dobra
Senegal	196,192 sq km 75,750 sq miles	Dakar	7,461,000	French	Islam, Christianity	Franc CFA
Seychelles	455 sq km 175 sq miles	Victoria	72,000	English, French, Creole	Christianity	Rupee
Sierra Leone	71,740 sq km 27,699 sq miles	Freetown	4,259,000	English	Traditional beliefs, Christianity	Leone
Tanzania	945,087 sq km 364,900 sq miles	Dodoma	28,342,000	Swahili, English	Christianity, Islam	Tanzanian shilling
Togo	56,785 sq km 21,925 sq miles	Lomé	3,563,000	French	Traditional beliefs, Christianity, Islam	Franc CFA
Uganda	235,880 sq km 91,074 sq miles	Kampala	19,095,000	Swahili, English	Christianity, Islam	Uganda shilling
Zaïre	2,344,885 sq km 905,365 sq miles	Kinshasa	37,145,000	French	Christianity	Zaïre

SOUTHERN AFRICA

Country	Area	Capital	Population	Language	Religion	Currency
Angola	1,246,700 sq km 481,354 sq miles	Luanda	10,002,000	Portuguese	Christianity, traditional beliefs	Kwanza
Botswana	581,730 sq km 224,607 sq miles	Gaborone	1,329,000	Tswana, English	Christianity, traditional beliefs	Pula
Comoros	2,235 sq km 863 sq miles	Moroni	535,000	Arabic, Swahili, French	Islam, Christianity	Franc CFA
Lesotho	30,355 sq km 11,720 sq miles	Maseru	1,824,000	Sesotho, English	Christianity, traditional beliefs	Loti
Madagascar	587,041 sq km 226,658 sq miles	Antananarivo	12,366,000	Malagasy, French	Christianity, Islam, traditional beliefs	Malagasy franc
Malawi	118,484 sq km 45,747 sq miles	Lilongwe	8,708,000	Chichewa, English	Christianity, traditional beliefs	Kwacha
Mauritius	2,040 sq km 788 sq miles	Port Louis	1,116,000	English, Creole	Hinduism, Christianity, Islam	Rupee
Mozambique	799,380 sq km 308,642 sq miles	Maputo	16,084,000	Portuguese, Bantu languages	Traditional beliefs Christianity, Islam	Metical
Namibia	823,145 sq km 317,818 sq miles	Windhoek	1,934,000	Afrikaans, English	Christianity	Rand
South Africa	1,221,037 sq km 471,445 sq miles	Cape Town, Pretoria	40,528,000	Afrikaans, English	Christianity, Hinduism, Islam	Rand
Swaziland	17,364 sq km 6,704 sq miles	Mbabane	816,000	English, Swazi	Christianity, traditional beliefs	Lilangeni
Zambia	752,614 sq km 290,586 sq miles	Lusaka	8,769,000	English	Christianity	Kwacha
Zimbabwe	390,580 sq km 150,804 sq miles	Harare	10,022,000	English	Christianity, traditional beliefs	Zimbabwe dollar

NOTE: The abbreviation CFA in 'franc CFA' stands for Communauté financiàre d'Afrique (African Financial Community); the franc CFA was a currency common to France's former colonies in Africa.

CANADA AND GREENLAND

Country	Area	Capital	Population	Language	Religion	Currency
Canada	9,970,610 sq km 3,849,744 sq miles	Ottawa	26,729,000	English, French	Christianity	Canadian dollar
Greenland	2,175,000 sq km 839,772 sq miles	Godthåb	57,000	Greenlandic, Danish	Christianity	Krone

NOTE: Greenland is a Danish province, with a large measure of self-government.

U.S.A

Country	Area	Capital	Population	Language	Religion	Currency
United States of America	9,372,571 sq km 3,618,770 sq miles	Washington DC	249,632,692	English, Spanish	Christianity, Islam, Judaism	US dollar

NOTE: Antarctica has no countries and no permanent inhabitants. Seven countries claim parts of Antarctica, but they have agreed that this barren land shall be used only for scientific research and exploration. The Falkland Islands, a South American British dependency, are the only inhabited lands of importance anywhere near Antarctica.

HIGHEST MOUNTAINS BY REGIONS

	m	ft
Central and Eastern Europe		
Mont Blanc (Alps-France)	4,807	15,771
Russia and Its Neighbours		
Elbrus (Caucasus-Georgia)	5,663	18,580
Central Asia		
Everest (Himalaya-Nepal, Tibet)	8,863	29,079
K2/Godwin-Austen (Pakistan, India)	8,611	28,250
Australia, New Zealand, and New Guinea		
Jaja (New Guinea)	5,029	16,500
Daam (New Guinea)	4,922	16,150
Antarctica		
Vinson Massif	5,140	16,864
North Africa		
Ras Dashan (Ethiopia)	4,620	15,158
Central Africa		
Kilimanjaro (Tanzania)	5,895	19,340
Southern Africa		
Thabana Ntlenyana (Drakensberg-Lesotho)	3,482	11,425
Alaska, Canada and Greenland		
McKinley (Alaska Range)	6,194	20,320
United States of America (excl. Alaska)		
Whitney	4,418	14,495
Tropical South America		
Huascaran (Peru)	6,768	22,205
Temperate South America		
Anaconagua (Andes-Argentina)	6,960	22,834

LARGEST DESERTS BY REGION

	sq km	sq miles
Central Asia		
Kara Kum (Turkmenistan)	270,000	105,100
Middle East		
Arabian Desert	1,300,000	500,000
Far East		
Gobi (China, Mongolia)	1,040,000	400,000
Indian Subcontinent		
Thar (India and Pakistan)	260,000	100,000
Australia, New Zealand and New Guinea		
Australian Desert	1,550,000	600,000
North Africa		
Sahara	8,400,000	3,250,000
Southern Africa		
Kalahari (Botswana)	520,000	200,000
United States of America		
Sonoran (USA and Mexico)	310,000	120,000
Temperate South America		
Atacama (Chile)	180,000	70,000

THE WORLD'S HIGHEST WATERFALLS

	m	ft
Angel (Venezuela)	979	3,212
Tugela (South Africa)	947	3,110
Utigård (Norway)	800	2,625
Mongefossen (Norway)	774	2,540
Yosemite (United States of America)	739	2,425
Mardalsfossen (Norway)	656	2,154
Tyssestrengane (Norway)	646	2,120
Cuquenan (Venezuela)	610	2,000
Sutherland (New Zealand)	580	1,904
Kjellfossen (Norway)	561	1,841

LONGEST RIVERS BY REGION

	km	miles
Northern Europe		
Shannon (Irish Republic)	386	240
Central and Eastern Europe		
Danube (Germany, Austria, Hungary, Serbia, Romania)	2,850	1,770
Elbe (Czech Republic, Germany)	1,160	720
Rhine (Switzerland, Germany, Netherlands)	1,320	820
Southern Europe		
Po (Italy)	660	410
Russia and Its Neighbours		
Amur (Siberia)	4,510	2,800
Dnepr (Russia, Belarus, Ukraine)	2,300	1,430
Ob'-Irtysh (Russia, Mongolia)	5,409	3,360
Volga (Russia)	3,531	2,195
Middle East		
Tigris-Euphrates (Turkey, Syria, Iraq)	2,800	1,740
Indian Subcontinent		
Brahmaputra (Tibet, India, Bangladesh)	2,900	1,800
Ganges (India, Bangladesh)	2,510	1,560
Indus (Tibet, Pakistan)	2,880	1,790
Far East		
Huang He (China)	4,672	2,900
Chang Jiang (China)	6,380	3,965
South-East Asia		
Mekong (Tibet, Thailand, Laos, Cambodia, Vietnam)	4,184	2,560
Australia, New Zealand and New Guinea		
Murray-Darling (Australia)	3,750	2,330
North Africa		
Nile (Kenya, Sudan, Egypt)	6,670	4,145
Central Africa		
Limpopo (South Africa, Mozambique)	1,610	1,000
Niger (Guinea, Mali, Niger, Nigeria)	4,184	2,600
Zaire or Congo	4,700	2,920
Zambezi (Zambia, Mozambique)	3,540	2,200
Southern Africa		
Orange (Lesotho, South Africa)	2,100	1,305
Alaska, Canada and Greenland		
Mackenzie-Peace (Canada)	4,240	2,635
St Lawrence (Canada, USA)	3,130	1,945
United States of America		
Colorado (also through Mexico)	2,334	1,450
Mississippi-Missouri	6,020	3,740
Tropical South America		
Amazon (Peru, Brazil)	6,440	4,000
Orinoco (Venezuela, Colombia)	2,740	1,700
Temperate South America		
Paraguay (Paraguay)	2,410	1,500
Rio de la Plata-Paraná (Brazil, Paraguay, Argentina)	4,880	3,030

THE WORLD'S LARGEST ISLANDS

	sq km	sq miles
Greenland	2,175,000	840,000
New Guinea	821,000	317,000
Borneo	744,360	287,400
Madagascar	587,040	226,600
Baffin Island (Canada)	507,450	195,900
Sumatra (Indonesia)	473,600	182,800
Honshu (Japan)	230,450	89,000
Great Britain	218,040	84,200
Victoria Island (Canada)	217,290	83,900
Ellesmere Island (Canada)	196,290	75,800

Acknowledgements

Photographs

Abbreviations: t = top; b = bottom; l = left; r = right; c = centre

Bryan & Cherry Alexander: 123t;
Biophoto Associates: 30r, 31r, 32r;
Catherine Blackie: 32t;
British Antarctic Survey/D. G. Allan: 68;
British Museum (Natural History): 30bl, 31c;
Department of Geology, University of Wales, Cardiff: 30tl, 31l, 32c;
Casella: 82-83;
J. Allan Cash: 104;
Bruce Coleman: 8-9 (Jen & Des Bartlett), 10 (Dieter & Mary Plage), 22 (William McPherson), 23 (Hans-Peter Merten), 24 (A. J. Deane), 35cr (G. Ziesler), 44 (John Shaw), 35bl (J. Fennell), 35br (Frieder Sauer), 45br (Leonard Lee Rue III), 46 (Gerald Cubitt), 49tr, 52t (Fritz Prenzel), 54 (David Coulston), 56 (L. G. Marigo), 65b (Adrian Davies), 73, 129 (Nicholas Devore III);
Colorific: 62-63t (Toby Sandford), 96 (Vadim Grippenreiter/ANA), 97t (John de Visser), 98-99t (Peter Turnley/Black Star), 98b (Emil Schulthers/Black Star), 99b (D. Wayman), 107b (Dallas & John Heaton), 124 (Mike Yamashita), 128 (Cary Wolinsky);
Robert Harding Picture Library: 16, 20, 21, 25, 40, 45bl, 50t, 57bl, 61, 86, 89b, 94, 100, 108-109b, 111t & b, 113, 114, 115, 116bl, 119cr, 132t & b, 133;
The Hutchison Library: 49cr, 52b, 79 (J. G. Fuller), 106 (Sarah Errington);
Impact Photos: 127r (John Cole);
Magnum: 89t (George Rodger), 90t (P. Zachmann), 91 (A. Venzago), 93l (Jean Gaumy), 97b (F. Mayer), 101b (Abbas), 105 (Abbas);
Marion & Tony Morrison, South American Pictures: 131, 134;
NASA: 6;
Novosti: 35t;
Oxford Scientific Films: 11 (Stan Osolinski), 13 (Andrew Plumptree), 19tl (Jack Wiburn), 19tr (G. A. Maclean), 32bl (Earth Sciences/Breck P. Kent), 47 (M. Wendler/Okapia), 48-49b (David Wrigglesworth), 66 (Richard Packwood), 67 (Paul Franklin), 69 (Michael Leach), 87t (Philippe Henry), 87b (Hjalmar R. Bardarson), 88 (Ronald Toms), 103 (Andy Park), 112l (Douglas Faulkner), 112br (Kjell Sandved), 118 (Steve Turner), 119bl (Owen Newman), 120t (Edward Parker), 121 (Richard Packwood), 125 (Stan Osolinski), 126-127b (H. Robinson), 130 (Andrew Plumptre);
Panos Pictures: 92t (J. Hartley), 108l (Chris Stowers), 109t (Mark McEvoy), 116-117t (Ron Giling), 117b (Neil Cooper), 120b (Ron Giling), 122t (David Reed), 122b (Trygve Bolstad);
Planet Earth Pictures: 19bl (John & Gillian Lythgoe), 28-29b, 53 (Ivor Edmonds);
Rex Features: 65t;
Science Photo Library: 82tl, 107t (Takeshi Takahara);
Tim Sharman: 90b, 92b;
Travel Photo International: 50b;
John Walmsley: 80l & r;
Zefa: 28tl, 38, 45tl, 45tr, 48, 57t, 57br, 58-59b, 59r, 60, 76, 93r, 95, 101t, 102, 110, 123b, 126tl.

Front cover:
top left: Oxford Scientific Films/Doug Allan
middle left: Oxford Scientific Films/Jason Rubinstein
middle right: Oxford Scientific Films/Kathie Atkinson
bottom left: Oxford Scientific Films/J. A. L. Cooke
bottom right: Oxford Scientific Films/Stan Osolinski

Back cover:
top left: Robert Harding Picture Library
middle left: Oxford Scientific Films/Okapia

Illustrations & diagrams

Gary Hincks: title page, 4, 6, 8-9, 12, 14-15, 17, 18, 26, 30, 31, 33, 42t, 72, 73, 74, 75r, 77; front cover bottom left, back cover bottom;
Peter Joyce: 62, 78;
Linden Artists (David Moore): 215;
Oxford Illustrators: 25, 29, 34, 38, 39, 64, 68, 70, 75l, 80, 81.

Index